D1772885

THE NUMBER-SYSTEM

THE NUMBER-SYSTEM

BY

H. A. THURSTON
M.A., Ph.D.
Department of Mathematics,
The University of British Columbia

DOVER PUBLICATIONS, INC.
NEW YORK

Copyright © 1956 by H. A. Thurston.
All rights reserved under Pan American and International Copyright Conventions.

Published in Canada by General Publishing Company, Ltd., 30 Lesmill Road, Don Mills, Toronto, Ontario.
Published in the United Kingdom by Constable and Company, Ltd., 10 Orange Street, London WC 2.

This Dover edition, first published in 1967, is an unabridged republication with minor corrections and revisions of the work originally published in 1956 by Blackie & Son, Ltd., and Interscience Publishers, Inc.

Standard Book Number: 486-61848-X
Library of Congress Catalog Card Number: 67-26869

Manufactured in the United States of America
Dover Publications, Inc.
180 Varick Street
New York, N.Y. 10014

PREFACE

IT IS NOT NECESSARY TO BE A WATCH-MAKER TO BE able to tell the time. Nor is it necessary to study the foundations of arithmetic in order to carry out the calculations required in everyday life, in science, or in mathematics. This book, however, is intended for those whose aim is to understand arithmetic rather than to put it to a practical use.

The systematic part of the book is the second part, and consists of Chapters **A-K**. It is complete in itself, and presents a detailed construction of the number-system, starting from Peano's axioms. Chapters **A-G** comprise a development of the rational number-system, and the detail is deliberately very full; every step, no matter how trivial, being justified by a reference to the theorem or lemma used. In Chapters **H-K**, first the theory of powers of rational numbers and then the real and complex number-systems are developed. Here the references are not quite so painstaking, the ordinary " laws of arithmetic " for rational numbers being freely used, as well as such familiar abbreviations as $x + y + z$ for $(x + y) + z$.

In order to keep the bulk of the second part of the book within bounds, explanatory matter has been cut to a minimum; but the first part of the book consists of explanatory chapters (I–XI) whose function is to explain and comment on the axioms and definitions, and to make the treatment in the second part digestible. The thoughtful non-mathematical reader who wants to know something about the number-system of modern mathematics without going deeply into the details could profitably read these chapters with perhaps a glance at the definitions and the statements of the theorems in the systematic part. But the reader for whom the book is primarily intended is the student of mathematics. He should read both parts, and probably the best order to take them in is as follows: I, II, **A**, III, IV, V, **B**, VI, **C**, **D**, **E**, **F**, **G**, VII, VIII, **H**, **I**, **J**, IX, **K**, X, XI.

The exercises are to help the reader to confirm that he has assimilated the ideas and to give him practice in using the concepts introduced, rather than to test his technique and industry. In general, solutions to the exercises should be strict; there is no point in proving that

$2 + 2 = 4$ unless this is properly done. But some illustrative exercises may be solved more freely. These are marked with asterisks.

A certain amount of pure algebra is introduced, partly for economy (the algebraic theorems being applied two or more times), partly to bring out the analogy between different parts of the arithmetical system, and partly to illustrate one use of abstract mathematics. The systematic chapters are for the most part alternately arithmetical and algebraic; the following chart shows their interdependence.

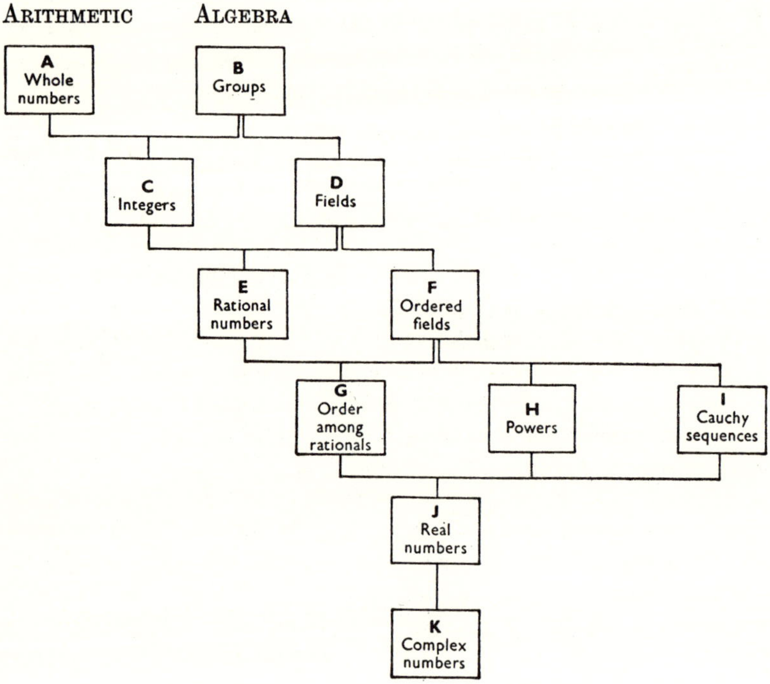

CONTENTS

PART I

EXPLANATORY TREATMENT

PART II

SYSTEMATIC TREATMENT

THE NUMBER-SYSTEM

Part I

EXPLANATORY TREATMENT

CHAPTER I

Counting

1. Numbers are used for counting; that is to say, for comparing the "many-fold-ness" of groups of objects. If we want to know whether there are more states in the United States than counties in England, the obvious thing to do is to count first the states and then the English counties. But the most obvious process is not in fact the simplest. We can see at a glance that there are more dots in [. . . . .] than in [. . .]. Some animals can distinguish between groups of different sizes (e.g. between a group of three men going into a hut, and a group of two of them coming out again), but there is no evidence that they can count.

The difference between counting and comparison without counting is brought out clearly if we imagine a primitive tribe whose language contains no numbers above 20. The tribesmen can then distinguish between groups of 17 and of 18 by counting, but not between groups of 25 and of 75. By the comparison method, on the other hand, the 25 and the 75 are far more easily distinguished than are the 17 and the 18. Comparison does not, however, depend on merely looking at the two groups. The crude method can be refined.

2. Let us imagine that a master of ceremonies wants to know whether there are more men or more women at a dance, and that he uses the comparison method, not just by glancing at the dancers but by asking everyone to take a partner. If there are men left over when all the women are partnered, then there are more men than women, and vice versa; if there is no one left over, then the sexes are present in equal strength. This process, simple though it is, is important in modern mathematics. I shall use the word *matching* to describe the pairing of the members of two sets,* and I shall call two sets *matchable* if they can be paired against each other with no members left over.

* We use the word *set* rather than *group*, because *group* has a special meaning in mathematics, whereas *set* is used quite generally. This is the reverse of ordinary usage, where, for example, a set of cigarette cards is a special group of cigarette cards: it contains just one card of each sort.

One set is of smaller size than another if and only if it is matchable with a set consisting of some members of the second set, but is not matchable with the second set itself.

3. A peasant farmer will perhaps be able to tell infallibly when one cow is more valuable than another. A more sophisticated merchant will not be content with this; he will want their values in terms of money. In other words, he wants a standard scale—the scale of money-values—instead of mere comparisons. We can construct a scale for our sizes-of-sets.

Consider [.], [. .], [. . .], [. . . .], and so on. If a set is matchable with the set of dots in [.], we say that the number of members of the set is *one*. If it is matchable with the set of dots in [. .], the number is *two*. Similarly for *three*, *four*, and higher numbers. We have now practically reached the stage of counting: counting is simply replacing the rather clumsy standard set by signs. The trick is to use a set of signs *arranged in a fixed order*. Let us suppose that we use the familiar signs 1, 2, 3, 4, etc. Then for our standard sets we take, not sets of dots, but sets of these signs, in the fixed order

[1] [1, 2] [1, 2, 3] and so on.

When we match a set of objects against one of the standard sets, we do not need to remember the whole standard set: it is enough to know which is the last number in it. Now at last we are really counting. We have used the fact that the set of standard sets is matchable with the set of number-signs, and have replaced each standard set by the number-sign with which it is paired. The matching can be illustrated thus:

[1] pairs with 1
[1, 2] pairs with 2
[1, 2, 3] pairs with 3
and so on.

Usually, matchings are written in the form

$$\begin{cases} [1] \longleftrightarrow 1 \\ [1, 2] \longleftrightarrow 2 \end{cases}$$

or by a general formula such as $[1, 2, \ldots, n] \longleftrightarrow n$.

1 and 2 are the *mates* of [1] and of [1, 2] respectively.

4. I want to emphasize that all we need for counting is a set of signs in a fixed order going on for ever. (Clearly the set must not ever come to an end: if it did we should not be able to count sets larger than the set of available signs.) The signs may be words:

one, two, three, four, ...

or

eins, zwei, drei, ...

or symbols:

1, 2, 3, 4, 5, 6, 7, 8, 9, 10, ...

or

. ⁚ ∴ ∷ ⁙ ⁝⁝ ⁘⁘ etc.

and so on.

That in our familiar system—the decimal system—we have only a small number of simple signs (1 to 9 and 0), using compound signs (10, 11, ..., 100, ...) for bigger numbers, is only a matter of convenience of writing. Calculation with these signs—the process taught at school as " arithmetic "—is really the study of the decimal notation rather than the study of *numbers*.

5. The reasoning in paragraph **3** shows that there is a close connection between counting and comparison without counting—so close that the reader may think that I am making too much fuss about the difference. But there is a reason. The process of pairing standard sets with number-signs arranged in order works only for finite numbers. Not only does the pairing process break down for infinite numbers, but it can be shown that the two ways of looking at numbers give two different systems of arithmetic, known as *cardinal* and *ordinal*. (This is not the same as the *grammatical* usage of the terms " cardinal " and " ordinal ". In grammar, the ordinal numbers are " first, second, third, ... ".)

6. " Infinite " (and similar words) are sometimes used in ordinary conversation to mean " very great ", and sometimes as synonyms for " unlimited ", " boundless ", etc. In certain other contexts their meaning is vague and mysterious: for example, " Of all the arts, dancing is perhaps the one most attuned to the infinite, having its essence in nature itself," Gopal and Dadachanji, *Indian Dancing*, p. 13. But in mathematics they are used in a perfectly definite sense. Quite simply, a set which has so many elements that the process of counting them one by one would never come to an end is said to be *infinite*, or to have an *infinite* number of (or an *infinity* of, or *infinitely* many) elements. To avoid confusion with the popular uses of the word, mathematicians often replace the term *infinite* by the less familiar word *transfinite*. An account of transfinite numbers will be found in J. E. Littlewood's *The Elements of the Theory of Real Functions*, or in Paul Halmos' *Naïve Set Theory*.

7. We have had to introduce the concept of " set ", and this concept is in fact a very important one in modern mathematics. It is even more fundamental than " number ", in spite of the fact that numbers are usually considered to be what mathematics is about.

The concepts of " set " and " property " are interchangeable inasmuch as we could use either concept in place of the other in any treatment of the subject. A given property $\boldsymbol{P}$, for example, determines a corresponding set: namely the set of things which have the property $\boldsymbol{P}$. And a set P determines a corresponding property: the property of belonging to P. Finally, if $\boldsymbol{P}$ corresponds to P in this way, then obviously P corresponds to $\boldsymbol{P}$. However, the mathematician prefers to work with sets, which seem to be a little more definite and amenable to calculation. For example, it is quite clear when two sets are the same: they are the same if they consist of the same elements; in other words, if every member of each belongs to the other. It is not quite so clear when properties are the same. As Bertrand Russell says in his *Introduction to Mathematical Philosophy* (p. 12), " Men may be defined as featherless bipeds, or as rational animals, or (more correctly) by the traits by which Swift delineates the Yahoos ". Does this mean that the property of being a featherless biped is the same as that of being a rational animal? Again, does the phrase " odd prime number " denote the same property as " prime number greater than 2 "? " Odd " is certainly not the same as " greater than 2 ". We might agree to consider two properties to be equal if and only if they define the same set; but this is precisely equivalent to working with sets instead of properties.

The concept of set is so important that there is a special notation for it: the expression $a \in A$ denotes that a belongs to the set A. Usually small letters will denote elements; large letters, sets. The symbol $\in$ will always denote " belongs to ".

CHAPTER II

Whole Numbers

1. If x and y are any two whole numbers, then $x + y = y + x$. In any given case we are able to verify this. For example, if we work out $5 + 3$ we get the answer 8. If we work out $3 + 5$ we get the same answer. But we cannot prove the general statement " if x and y are any two numbers, then $x + y = y + x$ " like this, because we cannot " work out " $x + y$. Nor can we prove it by testing all possible pairs of whole numbers, because there are an infinite number of them. And most people, when they try to prove the statement, find it extraordinarily difficult; not because it is too complicated but because it is too simple. To prove a complicated fact we use the simpler facts at our disposal. These simpler facts will usually have been proved from still simpler facts. And so on. Eventually we must get down to the simplest facts of all from which everything starts. These basic facts are called *axioms*.

The first thing to do, then, is to find a suitable set of axioms for an arithmetic of whole numbers. This simply means that from all the facts at our disposal we choose a few to form a foundation for our system—and all that is required of them is that they should suffice for this, i.e. that the whole system should be deducible from them. It is, of course, a practical advantage if they are fairly simple, and again it is convenient if they are self-evident, but neither of these properties is nowadays regarded as essential, though Euclid required his axioms to be self-evident. Usually there are any number of possible choices. To take a simple example, the laws of arithmetic (p. 12) could be taken as the axioms of a certain arithmetical system. If we were to replace the law $(x + y) \cdot z = x \cdot z + y \cdot z$ by the law $x \cdot (y + z) = x \cdot y + x \cdot z$, we should get a slightly different but equally good set of axioms for the same system—the *same* because clearly either law can be deduced from the other, using the commutative law for multiplication.

In order to know the simplest properties of whole numbers we must know what a whole number is, that is, we must consider not only axioms but also definitions. We shall first define *whole number*, *addition*, and *multiplication*. Subtraction and division are not so fundamental, and we may leave them until later.

2. We saw in Chapter I that the whole numbers are a set of symbols arranged in order: 1, 2, 3, . . . , for example. This statement is not quite definite enough for our purpose. Other things can be arranged in order: for instance, fractions can be arranged in ascending order of magnitude. But this is a different sort of order, because between any two fractions we can fit in others; for example, $\frac{1}{3}$, $\frac{1}{2}$, $\frac{2}{3}$, $\frac{4}{5}$ are in ascending order. But between $\frac{1}{3}$ and $\frac{1}{2}$ we can insert $\frac{5}{12}$ and so on. No fraction, in fact, has a next-in-order. This kind of order is of no use for counting. Having counted up to, say, 5, we must know that the next number is 6. Indeed, learning to count consists precisely of learning which number comes next after which. Thus we take as a fundamental property of whole numbers not just the fact that they are in order, but the fact that each has an immediate *successor*. An ordered set of this kind is called a *succession*. (The successor of a whole number is, of course, the number obtained by adding 1 to it; but we cannot *define* it like this, because we have not yet defined addition.)

We can now say exactly what we mean by whole numbers: they are the " descendants " of the first one. In other words, they consist of 1, the successor of 1, the successor of the successor of 1, and so on; and *only* of these. $2\frac{1}{2}$ is not a whole number: we cannot reach it by starting from 1 and travelling along the chain of successors. For the same reason, infinity is not a whole number.

This fact is expressed by the following two statements:

(*a*) 1 *is a whole number; and if x is a whole number, so is its successor.*
(*b*) *If a set of objects contains* 1 *and contains the successor of each of its members, then it contains all the whole numbers.*

(*a*) states that every successor of the successor of . . . the successor of 1 is a whole number; (*b*) that these are *all* the whole numbers.

As we write the numbers down in order, each one must be different from the ones already written. The set

$$1,\ 2,\ 3,\ 4,\ 5,\ 6,\ 7,\ 3,\ \ldots$$

would be useless, because 4 would have to follow 3, and the chain from 3 to 7 would repeat itself over and over again. Such a system would be a finite one, containing the numbers 1, 2, 3, 4, 5, 6, and 7 only. The difficulty arises because 2 and 7 have the same successor, 3.

We can overcome it by taking as an axiom:

No two different numbers have the same successor.

This axiom is not quite enough: it does not prevent the succession from going back to 1: e.g.

$$1, 2, 3, 4, 5, 1, 2, 3, \ldots$$

We can prevent this most simply by saying:

There is no number whose successor is 1.

We now have enough axioms for the whole of arithmetic. How do we know this? There is only one test: to deduce the whole of arithmetic from the axioms. This is done in Chapters **A–K**.

3. In elementary algebra a technique known as *proof by induction* is used. As an example, let us prove the formula

$$1 + 2 + 3 + \ldots + n = \tfrac{1}{2}n\cdot(n + 1)$$

with its help. First, we shall abbreviate the formula we are to prove. It is a statement involving the symbol n; let us denote it by $\boldsymbol{T}(n)$. Then $\boldsymbol{T}(3)$, for example, denotes the formula $1 + 2 + 3 = 6$; and $\boldsymbol{T}(n + 1)$ denotes

$$1 + 2 + \ldots + (n + 1) = \tfrac{1}{2}(n + 1)\cdot(n + 2)$$

We want to prove that the formula $\boldsymbol{T}(n)$ holds for all whole numbers n. Let us try to prove it step by step, proving $\boldsymbol{T}(1)$, $\boldsymbol{T}(2)$, $\boldsymbol{T}(3)$ and so on. After a little we find that it is natural to use the previous step in proving any given step. For example, in proving $\boldsymbol{T}(6)$ (having proved $\boldsymbol{T}(5)$, namely that $1 + 2 + 3 + 4 + 5 = 15$) we simply add 6 to $1 + 2 + 3 + 4 + 5$, which we know to be 15, instead of adding together 1, 2, 3, 4, 5, and 6. In other words, we let $\boldsymbol{T}(5)$ induce $\boldsymbol{T}(6)$. The principle of induction is simply this process stated in a more general way. We prove that $\boldsymbol{T}(n)$ induces $\boldsymbol{T}(n + 1)$ as follows.

If $\boldsymbol{T}(n)$ is true, then $1 + 2 + 3 + \ldots + n + (n + 1)$

$$= \tfrac{1}{2}n\cdot(n + 1) + (n + 1)$$

This we easily see to be equal to $\tfrac{1}{2}(n + 1)\cdot(n + 2)$, and so we have $\boldsymbol{T}(n + 1)$. This one formula ensures that $\boldsymbol{T}(1)$ induces $\boldsymbol{T}(2)$, $\boldsymbol{T}(2)$ induces $\boldsymbol{T}(3)$, $\boldsymbol{T}(3)$ induces $\boldsymbol{T}(4)$, and so on. We merely have to see that $\boldsymbol{T}(1)$ is true, and it follows that $\boldsymbol{T}(2)$ is true, then that $\boldsymbol{T}(3)$ is true, then that $\boldsymbol{T}(4)$ is true, and so on.

So much for elementary algebra. In the present treatment we have not defined addition, but we can use the idea of successor instead. Let us use a dash to denote a successor: r' is the successor of r. The principle of induction then becomes "If a statement $\boldsymbol{T}(1)$ is true and if $\boldsymbol{T}(r')$ is true whenever $\boldsymbol{T}(r)$ is true, then $\boldsymbol{T}(n)$ is true for every whole number n".

We can prove the principle from our axioms as follows. Let M be the set of all whole numbers for which $\boldsymbol{T}(n)$ is true. Then 1 belongs to M, because $\boldsymbol{T}(1)$ is true. Also, M contains the successor of each of its members, because if r belongs to M then $\boldsymbol{T}(r)$ is true, which implies that $\boldsymbol{T}(r')$ is true and therefore that r' belongs to M. Therefore, by axiom (*b*), M contains all the whole numbers, that is, $\boldsymbol{T}(n)$ is true for every whole number n.

4. Induction can be used for definitions as easily as for proofs. In elementary algebra the sign $n!$ (pronounced "factorial n") stands for the product $1 \times 2 \times 3 \times \ldots \times (n-1) \times n$. n can be any whole number from 1 upwards. But this formula is not a satisfactory definition. The three dots are not part of mathematical sign-language. The formula does indeed suggest what is meant, but is not good enough for a logical development. We can get rid of the dots, however, by defining $n!$ as follows:

$$\begin{cases} \text{(i)} \ 1! = 1 \\ \text{(ii)} \ n! = (n-1)! \times n \text{ if } n \text{ is a whole number greater than } 1 \end{cases}$$

Thus (i) defines 1! When 1! is defined, (ii) defines 2! When 2! is defined, (ii) defines 3! and so on. That "and so on" is the crux of the matter. Induction is a way of making an honest phrase of "and so on".

In elementary algebra induction is used for convenience; in our treatment of whole numbers we shall use it out of necessity. Axiom (*b*) is in effect an inductive definition of "all whole numbers", and so it is not surprising that if we want to say anything about "all whole numbers" we shall scarcely be able to avoid using the principle of induction.

Let us consider addition. How do we add 3 to 4? We count out the first three whole numbers, and then the next four; thus arriving at 7:

1	2	3	|	4	5	6	7
			|	1	2	3	4

By this simple process—which is that used by a child counting on his fingers—we can define addition by means of successors. To add 1, take the successor. To add 2, take the successor of the successor, and so on. We therefore make the following inductive definition of addition:

$$\begin{cases} \text{(i)} & x + 1 = x' \quad \text{for all } x \\ \text{(ii)} & x + y' = (x + y)' \quad \text{for all } x \text{ and } y \end{cases}$$

Thus, for each x, $x + 1$ is defined to be x' by (i). $2 = 1'$ and so, by (ii), $x + 2$ is defined to be $(x + 1)'$, that is, x''. Then $x + 3$ is x''', and so on.

Similar arguments apply to multiplication. Multiplication is essentially repeated addition. " Three fours " is $4 + 4 + 4$. Thus, having multiplied m by n, to multiply m by $n + 1$ we need only add an extra m. This gives us the following definition of multiplication:

$$\begin{cases} (a) & x \cdot 1 = x \\ (b) & x \cdot y' = x \cdot y + x \end{cases}$$

From these two definitions and our axioms we can now develop the complete theory of whole numbers. In particular we can prove the laws of arithmetic, one of which—$x + y = y + x$—has already been mentioned. This is done in Chapter **A**, pp. 51–58.

A system of numbers with operations such as addition and multiplication in which calculations can be carried out is an *arithmetic*. Thus in Chapter **A** we develop the arithmetic of whole numbers. We shall make one slight alteration in our treatment there: we shall start from 0 instead of from 1, because it is convenient to count 0 as a whole number. From now on we shall assume all the results of this arithmetic.

5. *Note.*—Readers who are used to induction in elementary algebra may have to be careful about one point. One form of the principle runs " If $\boldsymbol{T}(1)$ is true, and if we can prove $\boldsymbol{T}(r)$ by assuming that $\boldsymbol{T}(s)$ is true whenever s is less than r; then $\boldsymbol{T}(n)$ is true for all whole numbers n ". But we cannot use this form of the principle yet, because our axioms say nothing about the concept of " less than ". The inductive definitions in **A3** (p. 51) and **A9** (p. 54) are proved to be valid without the use of the order-relation; but the inductive definition of *power* in **H2** (p. 91) comes after the definition of this concept, and so the second form of the principle can be used.

CHAPTER III

The Laws of Arithmetic

1. One often has to add more than two numbers together: to evaluate $3 + 4 + 7$ for example. What is meant by this? We have defined $x + y$, but not $x + y + z$. No doubt we could define $x + y + z$, and $x + y + z + w$, and so on, in a similar though more complicated way, but fortunately there is no need. We can add 3 to 4 and then add the result to 7. We could have added 3 to the result of adding 4 to 7. And anyone confronted with the formula $3 + 4 + 7$ will probably add together two of the numbers and then add the third to the result, though he will probably do it so quickly that he does not see the two separate steps. Thus we have no need of an expression like $x + y + z$: the expressions $(x + y) + z$ and $x + (y + z)$ will do instead, but—and this is the important point—only because they are equal. Several of the more important facts such as $(x + y) + z = x + (y + z)$ and $x + y = y + x$ are singled out for special attention and called *laws of arithmetic*. They are shown in the table.

	Addition	*Multiplication*
Commutative law	$x + y = y + x$	$x \cdot y = y \cdot x$
Associative law	$(x + y) + z = x + (y + z)$	$(x \cdot y) \cdot z = x \cdot (y \cdot z)$
Cancellation law	If $x + y = x + z$, then $y = z$	If $x \cdot y = x \cdot z$ and $x \neq 0$. then $y = z$
Distributive law	$(x + y) \cdot z = x \cdot z + y \cdot z$	
Neutrality of 0* and 1	$0 + x = x$	$1 \cdot x = x$

* Some writers count 0 as a whole number, others do not. If we do not, then no whole number is neutral for addition, and the proviso that $x \neq 0$ in the cancellation law of multiplication is unnecessary. In the first part of the book we shall sometimes include and sometimes exclude 0, as convenient for different purposes. In the systematic part of the book we shall, of course, be systematic; there the whole numbers will definitely include 0.

These are all proved, when x, y, and z are whole numbers, in Chapter **A.** These laws are important, and not only because they are useful for carrying out calculations. It is possible to prove many of the interesting properties of numbers from them alone, without referring back to the axioms. Such formulæ as $(x+y)+z=(z+y)+x$ and $(x+y)+(z+w)=(x+z)+(y+w)$ follow straight from the commutative and associative laws for addition.

As a slightly harder example, let us prove from the laws of arithmetic only that $0 \cdot x = 0$.

$$y \cdot x + 0 \cdot x = (y+0) \cdot x \quad \text{by the distributive law}$$
$$= y \cdot x \quad \text{by the neutrality of 0 for addition}$$
$$= y \cdot x + 0 \quad \text{by the neutrality of 0 for addition.}$$

Therefore $0 \cdot x = 0$ by the cancellation law for addition.

2. We can now look at our system from another point of view. It consists of (i) a set of elements, namely the numbers 0, 1, 2, . . . , and (ii) a set of operations, namely addition and multiplication. And we could in fact have defined the arithmetic of whole numbers to be the set $\{0, 1, 2, \ldots\}$ with the operations of addition and multiplication obeying the laws of arithmetic, instead of defining it in terms of the axioms. We have now entered the realm of modern algebra.

An *algebraic system* consists of a set of elements with a set of operations obeying certain laws. An operation can be formally defined as follows: it is a rule whereby to each ordered * pair of elements of the set there corresponds a third element of the set. (The third element corresponding to the ordered pair (x, y) is $x+y$ if the operation is addition, $x \cdot y$ if the operation is multiplication.) According to the particular laws obeyed, so we get different types of algebraic system.

Algebraic systems are *abstract*: the elements may be any objects whatever, and the algebraist is interested only in the structure of the system as given by the operations. An algebraic system can be constructed quite readily by taking any set of elements, writing out a double-entry table, and filling in the cells arbitrarily. We can define an operation $*$ from the table by letting $x * y$ be the entry in row x column y. Making such a table for each operation, we have an algebraic system. The diagram depicts a system with three elements and one operation.

	a	*b*	*c*
a	*a*	*c*	*b*
b	*b*	*a*	*c*
c	*c*	*b*	*a*

* A pair is ordered if it matters which of the two objects is considered first. "Ordered" corresponds to "respectively". Thus (m, n) is a different *ordered pair* from (n, m), unless $m = n$. But "m and n" is the same *pair* as "n and m".

For example, $b * a = b$, and $a * b = c$. Thus this particular operation does not obey the commutative law (which is that $x * y = y * x$). The reader may like to consider the other laws of arithmetic.*

More often we do not confine our attention to a single system, but consider a whole class of systems defined in the following way: we specify the number of operations and certain laws which the operations are to obey. Systems satisfying a particular definition are given the appropriate name (group, field, etc.) and the laws which form part of the definition are often called axioms with the appropriate prefix. Thus laws (i)–(iv) below could be referred to collectively as the *hemigroup axioms*.

To see whether a given system belongs to a given class of systems we have simply to check that it has the right number of operations and that the axioms are obeyed.

3. The simplest algebraic systems are those with only one operation. Sometimes this is written as addition or as multiplication—a mere matter of notation. If the operation is written as addition the algebra is termed *additive*. The language appropriate to addition will, of course, be used. For example, $a + b$ will be called the *sum* of a and b even when $+$ represents an abstract operation additively written and not addition in the usual sense.

Two types of algebraic system are worth defining: first, a *hemigroup*. A hemigroup is a set of elements with one operation, $*$, obeying the following laws.

(i) $(x * y) * z = x * (y * z)$ for every x, y, and z.

(ii) $x * y = y * x$ for every x and y.

(iii) If $x * y = x * z$, then $y = z$.

(iv) There is an element e such that $e * e = e$.

The whole numbers (including 0) with the operation of addition form a hemigroup, 0 playing the part of e. So do the whole numbers (without 0) with the operation of multiplication, 1 playing the part of e. The whole numbers including 0 do not form a hemigroup under multiplication: the third law is not satisfied, because $0 \cdot 1 = 0 \cdot 2$ although $1 \neq 2$.

The second important type of algebraic system is the *commutative*

* $(b * b) * b = a * b = c$. $b * (b * b) = b * a = b$. Therefore the operation does not obey the associative law. a is a right-neutral element ($x * a = x$ for every x) but there is no left-neutral element. The left- and right-cancellation laws are obeyed.

group. A commutative group is a set of elements with one operation obeying (i)–(iii) and also

(v) Given any a and b there is a x such that $a * x = b$.

Neither of the hemigroups mentioned above is a commutative group: there is no whole number x for which $x + 1 = 0$, and there is no whole number x for which $2 \cdot x = 1$. But, to anticipate the introduction of negative numbers and fractions, two important commutative groups are (i) the integers (i.e. the positive * and negative whole numbers) under addition, and (ii) the positive non-zero fractions under multiplication. (The solution of $a + x = b$ is $b - a$, which is an integer if a and b are; the solution of $a \cdot x = b$ is b/a, which is a positive non-zero fraction if a and b are.)

Although algebra can be usefully applied to arithmetic, its application is by no means confined to number-systems. A delightfully simple example of a commutative group whose elements are not numbers has been described by Professor M. H. A. Newman: the elements are two actions—*crossing the road* and *not crossing the road.* The product of two actions is the action which has the same result as performing one action and then the other. If we denote the actions by s and t respectively, we get $s * s = t * t = t$ and $s * t = t * s = s$. (For example, crossing the road and then crossing it again brings the pedestrian back to the same side and so is equivalent to not crossing the road; i.e. $s * s = t$.) It is easy to verify that the relevant axioms are obeyed.

4. One consequence of the fact that an abstract algebraist is interested only in the *structure* of a system is that two systems are considered equivalent if one can be derived from the other by a change of notation. Suppose, for example, that we construct a system having three elements f, g, h and one operation, $\circ$, given by the table

$\circ$	f	g	h
f	f	h	g
g	g	f	h
h	h	g	f

This can be derived from the system defined in § 2 by writing f in place of a, g in place of b, h in place of c, and $\circ$ in place of $*$. In other

* The word "positive" is used by different writers in two different ways: it can either include or exclude zero. I shall use it in the admittedly less common inclusive sense—the number 0 is both positive and negative. (In the other sense, 0 is neither positive nor negative.) The proposed usage agrees with that in N. Bourbaki's *Eléments de Mathématique*, Hermann, Paris, but disagrees with traditional usage. The traditional term corresponding to Bourbaki's "positive" is the rather clumsy "non-negative".

words, we have a matching

$$\begin{aligned} a &\longleftrightarrow f \\ b &\longleftrightarrow g \\ c &\longleftrightarrow h \end{aligned}$$

between the two sets of elements, and a matching

$$* \longleftrightarrow \circ$$

between the operations. And, finally, when we apply the matchings to the table for one operation, we get the table for the other. For example, the c in the top row of the first table indicates that $a * b = c$. Applying the matchings, this equation becomes $f \circ g = h$. But $c \longleftrightarrow h$ in the first matching, which means that $a * b \longleftrightarrow f \circ g$. And in general, we see that a matching is equivalent to a change of notation precisely when the following is true:

$$x \longleftrightarrow u \text{ and } y \longleftrightarrow v \text{ always imply } x * y \longleftrightarrow u \circ v$$

Often the same sign is used for both operations. If the sign is $*$, a matching with this property is called a $*$-*isomorphism.* If the system has several operations, we shall want the matching to have the corresponding property for each of them. If it has, it is called simply an *isomorphism.* Two systems between which there is an isomorphism are *isomorphic.* For example, the system whose elements are 1 and —1 and whose only operation is ordinary multiplication; the system whose elements are 0 and 1 and whose only operation is addition *modulo* 2;* the system given by the following table:

$*$	e	f
e	e	f
f	f	e

and Professor Newman's " crossing the road " group are all isomorphic to one another.

* Addition *modulo* 2 simply means ordinary addition followed by the operation of taking the remainder when the sum is divided by 2. For example, $1\frac{1}{2} + 1\frac{3}{4}$ *modulo* 2 is equal to $1\frac{1}{4}$. The relevant facts for our example are that

$$\left.\begin{array}{l} 1 + 0 = 0 + 1 = 1 \\ 0 + 0 = 1 + 1 = 0 \end{array}\right\} \textit{modulo } 2$$

CHAPTER IV

Fractions

1. The whole numbers are adequate for counting the change when shopping and for similar everyday calculations. But for other calculations they are inadequate: we want perhaps to weigh parts of a pound or to measure fractions of a foot. What number is suitable for measuring the weight of half of a five-pound cake? If x is the number of pounds in the half-cake, we must have $2 \cdot x = 5$. But there is no whole number for which this is true. The equation $a \cdot x = b$ is not always solvable. We must therefore invent extra numbers—fractions—to make this equation solvable as far as possible.

Without going into the logical development of mathematics, every schoolboy learns about fractions. He knows, for example, that

$$(a/b) \cdot (c/d) = (a \cdot c)/(b \cdot d),$$

and that

$$(a/b) + (c/d) = (a \cdot d + b \cdot c)/(b \cdot d)$$

It is worth while pausing to consider how he knows these facts.

"Cut a cake into d equal parts and take c of them. You now have c $d$ths of the cake, represented by the fraction c/d. Divide this into b equal parts and take a of them. You now have a $b$ths of c $d$ths of the cake, represented by the expression $(a/b) \cdot (c/d)$. The obvious way to take the second step is to divide *each* of the $d$ths into b equal parts and take a of them. The total number of small parts is then $a \cdot c$, and each is a $(b \cdot d)$th of the cake. You therefore have $a \cdot c$ $(b \cdot d)$ths of the cake, and so $(a/b) \cdot (c/d) = (a \cdot c)/(b \cdot d)$." An argument like this is sometimes given as a proof of the formula. It is in fact nothing of the sort. It is a statement not about numbers but about weights of pieces of cake. Numbers are used for other things besides weights—lengths, for example—and no amount of argument about weights will prove that half of half an inch is a quarter of an inch.

How then, can we prove our formulæ? We must first decide what we mean by addition and multiplication of fractions. When we have defined these concepts, we can hope to prove our formulæ from the definitions; and before we do this, we must define *fraction*. We shall assume all the properties of whole numbers, and from them define or deduce the properties of fractions. In other words: starting from the arithmetic of whole numbers, we shall construct an arithmetic of fractions. Then arguments like the one given above, which are not proofs of anything about fractions, become verifications that the arithmetic of fractions is suitable for dealing with weights, measures, and so on.

The question whether a given problem can be dealt with by whatever arithmetic we have at our disposal is an important one. The answer to the question "If a bun costs 1*d.* how much do 27 buns cost?" is 2*s.* 3*d.*, provided that the arithmetic of whole numbers is applicable. If the baker sells 13 for 1*s.* (as many do), the arithmetic is not applicable; and the answer is 2*s.* 1*d.* In practice we use our common sense in deciding whether to apply the usual arithmetic.

2. The details will be simpler (and the example just as informative) if we start from the *non-zero* whole numbers and construct the non-zero fractions; and this is what we shall do.

We want to make the equation $a \cdot x = b$ solvable by introducing extra numbers, which we are going to call fractions, and we want the laws of arithmetic to be true for them. We shall find that this condition will be a sufficient guide.

First, it follows from the cancellation law that the equation cannot have more than one solution, for if p and q are solutions, then $a \cdot p = b = a \cdot q$ and so, by the cancellation law, $p = q$. The solution of $a \cdot x = b$ depends on a and b; we shall therefore denote it by a symbol involving a and b, namely by b/a.

Sometimes two different equations have the same solution; $2 \cdot x = 12$ and $3 \cdot x = 18$ have both the solution $x = 6$, for example. Just when does this happen? Suppose that $a \cdot x = b$ and $c \cdot x = d$ have the same solution, $x = k$. Then $a \cdot k = b$ and $c \cdot k = d$. Therefore $(a \cdot k) \cdot d = b \cdot (c \cdot k)$. Therefore, if the commutative and associative laws are to hold, we must have $(a \cdot d) \cdot k = (b \cdot c) \cdot k$. Then the cancellation law tells us that $a \cdot d = b \cdot c$. Suppose, on the other hand, that a, b, c, and d are numbers such that $a \cdot d = b \cdot c$. Let k be a solution of $a \cdot x = b$. Then $a \cdot k = b$. Therefore $(a \cdot k) \cdot c = b \cdot c = a \cdot d$. Therefore by the associative law, $a \cdot (k \cdot c) = a \cdot d$. Then, by the cancellation and commutative laws, $c \cdot k = d$, and so k is a solution of $c \cdot x = d$. To say that k is a solution of $a \cdot x = b$ is the same thing as saying that $k = b/a$.

Therefore we have proved that if the laws of arithmetic are to hold, we must have equality between fractions as follows:

$$b/a = d/c \text{ if and only if } a{\cdot}d = b{\cdot}c$$

Now we can start on our definition: A fraction is something represented by a symbol of the form a/b where a and b are any whole numbers.* a/b and c/d are equal if and only if $a{\cdot}d = b{\cdot}c$.

Let us consider addition. How must we define $(a/b) + (c/d)$ if the laws of arithmetic are to be true? If the distributive law is true, then

$$(b{\cdot}d){\cdot}((a/b) + c/d)) = (b{\cdot}d){\cdot}(a/b) + (b{\cdot}d){\cdot}(c/d)$$

If the associative and commutative laws are true, this is equal to

(I) $$d{\cdot}(b{\cdot}(a/b)) + b{\cdot}(d{\cdot}(c/d))$$

Now, by definition, a/b is the solution of $b{\cdot}x = a$: that is, $b{\cdot}(a/b) = a$. Similarly, $d{\cdot}(c/d) = c$. Therefore (I) is equal to $d{\cdot}a + b{\cdot}c$. Therefore $(a/b) + (c/d)$ is the solution of $(b{\cdot}d){\cdot}x = d{\cdot}a + b{\cdot}c$; that is, $(a/b) + (c/d)$ is $(d{\cdot}a + b{\cdot}c)/(b{\cdot}d)$.

This, then, must be our definition. Similarly we can show that our definition of $(a/b){\cdot}(c/d)$ must be $(a{\cdot}c)/(b{\cdot}d)$.

Notice that these definitions do in fact define the sum and product of fractions as fractions. For example, $d{\cdot}a + b{\cdot}c$ is a whole number if a, b, c, and d are. And $b{\cdot}d$ is a (non-zero) whole number if b and d are. Therefore $(d{\cdot}a + b{\cdot}c)/(b{\cdot}d)$ is a fraction (unless $d{\cdot}a + b{\cdot}c$ is zero). Similarly for the product.

3. So far we have been arguing backwards—the definitions must be like this if we are to get the properties we want. Now we must argue forwards: we must make the above definitions and from them prove the properties, including all the laws of arithmetic for fractions. The details of this will be found in the systematic chapters. But first there are one or two difficulties to be overcome.

A careful logically-minded reader may complain, "It is all very well to say 'the objects we are going to deal with are represented by symbols of the form a/b' but what are the objects? And how do we know that there are any objects representable in the way you describe?" The objection is valid. It is easy to run into trouble in mathematics by blithely assuming the existence of objects which we want to exist. This particular objection, however, can be met by de-

* Remember that we are not counting 0 as a whole number.

fining the objects, though they are rather abstract and unlike what we usually imagine fractions to be. (That is why we have separated out the definition in this way.)

We pointed out earlier (Chapter II, § **1**) that in an investigation such as the one we are engaged on, we must start from a system of basic axioms and definitions, and build our mathematics upon them. We have now reached a point where we must consider carefully just what our basic concepts shall be. This book is about some of the simplest things in mathematics, the whole numbers. Our basic concepts must be something simpler still. I use the word " simpler " in a technical sense; the reader may not find the simpler concepts easier or more natural. Our basic concepts must in fact be logically prior to mathematics; they must be logical and linguistic concepts.

Such concepts are those denoted by the terms " and ", " or ", " not ", " implies " (or some synonym or paraphrase), and " statement ". The reader will be in no doubt about the meaning and usage of these (though it may be as well to emphasize that in mathematics " x or y " always means " x or y or both ", never " x or y but not both "). Another fundamental concept is that of " set " and the terms that go with it: " contains ", " belongs to ", " member ", and so on.

It may be convenient for the reader to consider " relation " and " function " as two other basic words, though they can in fact be defined in terms of " set ".

4. Now let us return to the definition of " fraction ". Assuming the existence of whole numbers, there is no doubt of the existence of symbols like m/n, each obtained by writing a whole number, a stroke, and then another whole number. They will not, however, do for fractions because the symbol 1/2, for instance, is different from the symbol 2/4, whereas we want these two fractions to be equal.

Let us consider all those ordered pairs (m, n) of whole numbers which have the property that $n = 2{\cdot}m$. We take this set of ordered pairs as one of our objects, and call it the fraction 1/2. Similarly, if (x, y) is any ordered pair of whole numbers, we define the fraction x/y to be the set of all ordered pairs (m, n) for which $m{\cdot}y = n{\cdot}x$. Thus a fraction is a set of ordered pairs of whole numbers.

Now let us prove that $1/2 = 2/4$. To prove that two sets of objects are identical we have to show that every object in the first set is in the second and vice versa.

The condition for (m, n) to be in the fraction 1/2 is $n = 2{\cdot}m$. The condition for (m, n) to be in the fraction 2/4 is $2{\cdot}n = 4{\cdot}m$. It is one of the properties of the arithmetic of whole numbers that $n = 2{\cdot}m$ if and only if $2{\cdot}n = 4{\cdot}m$. Therefore an ordered pair is in one fraction if and only if it is in the other, and so the two fractions are equal. In this way we can show, in general, that a/b and c/d are equal if and only if $a{\cdot}d = b{\cdot}c$. This is precisely what we wanted.

5. In everyday language we use fractions and whole numbers together: when we ask for a three-halfpenny stamp we are using the fraction 3/2; when we ask for a penny-halfpenny stamp we are using the mixed number $1\frac{1}{2}$. Therefore it would be an advantage to have the whole numbers included in our arithmetic of fractions. But, on our definition, fractions are quite different from whole numbers: they are sets of pairs of whole numbers. Luckily, however, our fractions include something just as good: namely, the fractions of the form $x/1$. We might call these the " whole fractions ".

When we say that they are " just as good " there are two things we might mean, according to whether we take the axiomatic or the algebraic viewpoint. Axiomatically, we mean simply that they obey the axioms of the arithmetic of whole numbers. We take 1/1 for our starting-point and define the successor of $x/1$ to be $x'/1$, where x' is the successor of x. The axioms are then quite easy to verify.

Algebraically, we want to show that the whole fractions form a system which is isomorphic to the whole numbers. The multiplication tables for the two systems are:

	1	2	3	4 . . .
1	1	2	3	4
2	2	4	6	8
3	3	6	9	12

	1/1	2/1	3/1	4/1 . . .
1/1	1/1	2/1	3/1	4/1
2/1	2/1	4/1	6/1	8/1
3/1	3/1	6/1	9/1	12/1

We can turn the first into the second by writing $x/1$ in place of x everywhere. Thus the matching $x \longleftrightarrow x/1$ is a multiplication-isomorphism. Equally clearly we see that it is an addition-isomorphism.

We shall use the idea of isomorphism over and over again in the same way. We shall be continually enlarging our number-system. We shall not want to lose sight of the simpler systems we started from, so we shall pick out every time part of the new system which is isomorphic to the old. Thus none of our work is lost; all the results we prove about earlier systems will be true for certain parts of each later one. A system which is isomorphic to part of another is said to be *embedded* in the larger system. We shall see that the process of defining the number-system, once the whole numbers are defined, consists of a series of embeddings. The whole numbers are embedded in the integers; the integers, in the rational numbers; the rational numbers, in the real numbers; and the real numbers, in the complex numbers.

CHAPTER V

Negative Numbers

1. In ordinary life, we are satisfied with whole numbers and fractions. But in quite simple mathematics these prove insufficient. Coördinate geometry would be very clumsy if we adopted the geographers' system and wrote coördinates as 4° North and 5° East, 4° North and 5° West, instead of (4, 5), (4, —5), and so on. If a pure mathematician worked in a bank, he would probably not refer to credit balances and debit balances: he would simply refer to balances, and represent debit balances by negative numbers. This is in fact the way in which negative numbers most naturally arise. If £5 is withdrawn from a balance of £3, the result is a debit balance of £2. In the arithmetic of whole numbers, 5 cannot be subtracted from 3. We therefore construct new numbers which will permit this, that is, we make the equation $a + x = b$ always solvable.

2. We proceed in the same way as when we constructed numbers to solve $a \cdot x = b$. If we start from the whole numbers, we get the integers; if we start from the fractions, we get the positive and negative fractions, usually called rational numbers.

The solution of $a + x = b$ depends on a and b: denote it by $b - a$. Then $a + (b - a) = b$.

How shall we define equality? Assume that the laws of arithmetic hold and that $b - a = d - c$. Then $b + c = a + (b - a) + c = a + (d - c) + c = a + d$. Conversely, if $b + c = a + d$, then $a + (b - a) + c = b + c = a + d$, and so $(b - a) + c = d$. Therefore $c + (b - a) = d$; that is, $b - a = d - c$. Therefore we want $b - a$ and $d - c$ to be equal if and only if $b + c = a + d$.

How shall we define addition? If the associative and commutative laws are true,

$$(a + c) + ((b - a) + (d - c)) = (a + (b - a)) + (c + (d - c)),$$

and this is equal to $b + d$, by the definition of $b - a$ and $d - c$. Therefore, again by definition, $(b - a) + (d - c)$ is $(b + d) - (a + c)$.

Similarly we can prove that the product $(b - a)\cdot(d - c)$ can only be $(a\cdot c + b\cdot d) - (b\cdot c + a\cdot d)$.

Finally, we can make $a - b$ satisfy our desired conditions for equality, by defining it to be the set of all ordered pairs (m, n) for which $a + n = b + m$.

3. A close analogy between this process and the one described in Chapter IV is now clear. First, let us notice that our present object —to solve $a + x = b$—can be attained without any mention of multiplication (though we cannot define multiplication of our new objects without using multiplication of the numbers we start from). Similarly, fractions can be defined, and multiplication of fractions also, without reference to addition. Let us consider the two procedures side by side.

We start from a system in which

(i)	$x + y = y + x$	$x\cdot y = y\cdot x$
(ii)	$(x + y) + z = x + (y + z)$	$(x\cdot y)\cdot z = x\cdot(y\cdot z)$
(iii)	If $x + y = z + y$, then $x = z$	If $x\cdot y = z\cdot y$, then $x = z$
(iv)	$0 + 0 = 0$	$1\cdot 1 = 1$
(v)	$a + x = b$ is not always solvable	$a\cdot x = b$ is not always solvable

and create a system in which (i), (ii), (iii), and (iv) are still true and the equation in (v) is always solvable, and which contains a subsystem isomorphic to the system from which we started. Our procedure is to make the definitions

(*a*)	$a - b$ is the set of (m, n) for which $a + n = b + m$	a/b is the set of (m, n) for which $a\cdot n = b\cdot m$
(*b*)	$(a - b) + (c - d)$ is $(a + c) - (b + d)$	$(a/b)\cdot(c/d)$ is $(a\cdot c)/(b\cdot d)$

The difference is merely one of notation. If we replace sums by products, the statements in the first column become the corresponding statements in the second. The two procedures are, in fact, both special cases of the following: if we start from a hemigroup H (whose operation we shall denote by $*$) and define new objects $a \mid b$ and an operation $*$ on them by

(*a*) $a \mid b$ is the set of (m, n) for which $a * n = b * m$

(*b*) $(a \mid b) * (c \mid d)$ is $(a * c) \mid (b * d)$

then the resulting system is a commutative group and contains a hemigroup $*$-isomorphic to H. The hemigroup H is embedded in the group.

4. In the systematic development we shall not do the same work twice, but shall devote Chapter **B** to embedding a hemigroup in a commutative group. In Chapter **C** we shall apply this to the whole numbers as an additive hemigroup, and shall be able to write down the additive properties of the integers straight away. We complete the arithmetic of integers by defining multiplication as in Chapter IV. In Chapter **E** we apply Chapter **B** again, this time to the non-zero integers (which are a multiplicative hemigroup), and so can write down all the multiplicative properties. We complete the arithmetic of rational numbers by defining addition.

5. This is a good example of the technique of abstract mathematics. For thousands of years mathematicians have examined the properties of certain things which are susceptible to quantitative investigation: numbers, for example, (arithmetic and analysis) or space (geometry and trigonometry). The more careful treatments have started from axioms, which are simply a few fundamental properties from which all the others can be shown to follow. Abstract mathematics has its axioms too, but they are looked on in a different way: they are part of the definitions of the systems involved. (i)–(iv) are the axioms of a hemigroup. They resemble the classical axioms in that all the properties of the system under discussion follow from them. They differ in that the question of their being true or false does not arise. Every theorem in abstract algebra is of the form, " If a system obeys such-and-such axioms, then it has such-and-such properties ". If the system did not obey the axioms, then the theorem would still be true, but the conclusion could not be asserted.

Now if we have a system for which certain axioms are true, we can then assert all the conclusions deduced by abstract reasoning from these axioms. If whole numbers form a hemigroup (as in fact they do) we can say that the conclusion of any theorem proved about hemigroups is true for the arithmetic of whole numbers. The system for which the axioms of the abstract system are true is then a *realization* or *model* of the abstract system. An abstract algebraic system has been likened to an electric vacuum-cleaner which can be plugged in at many convenient points. We plug the hemigroup in at two points: Chapter **C** and Chapter **E**.

CHAPTER VI

Fields

1. One of the most important abstract algebraic systems is the field. This is a system which is doubly a group in the following way: first, it is a commutative group under an operation which we shall denote by $+$ and call addition. We shall denote the neutral element of this group by 0 and call it zero. Second, the non-zero elements of the system form a commutative group under a second operation, which we call multiplication. Finally, the distributive law holds: $x \cdot (y + z) = x \cdot y + x \cdot z$. The distributive law is the only one which involves both addition and multiplication. Without it, the two groups are coëxistent but uncoördinated. The distributive law makes their union fertile.

2. In a field we can define subtraction and division: $b - a$ is the x for which $a + x = b$, and is defined for every a and b; d/c is the y for which $c \cdot y = d$, and is defined for every d and every non-zero c. We can define subtraction and division in systems which are not fields: if a and b are whole numbers, $b - a$ is the whole number x for which $a + x = b$. It exists if and only if b is greater than a. Such a definition would not be convenient from an algebraist's viewpoint; too many statements would have to be qualified by such provisos as " if it exists ".

Subtraction and division are derived from addition and multiplication respectively. It is therefore to be expected that their chief properties will be easily deducible from those of the primary operations. As an example, let us prove the distributive law of multiplication and subtraction, namely that $x \cdot (y - z) = x \cdot y - x \cdot z$.

By definition, $y - z$ is the number u for which $z + u = y$. By the distributive law of multiplication and addition,

$$x \cdot y = x \cdot (z + u) = x \cdot z + x \cdot u$$

That is, $x \cdot u = x \cdot y - x \cdot z$. Therefore $x \cdot (y - z) = x \cdot y - x \cdot z$.

The reader may, as an exercise, like to prove the distributive law of division and addition, namely,

$$(x + y)/z = (x/z) + (y/z)$$

This holds, of course, for every x, every y, and every non-zero z.

The use of the notation $b - a$ (the same notation as we used for embedding an additive hemigroup in a group) is deliberate. The two concepts are analogous. The difference is that in the present case $b - a$ is an element of the system we are considering, but in the former case it is something specially created in order to solve the equation $a + x = b$.

3. A field can justly be described as an algebraic system in which addition, multiplication, subtraction, and division can be carried out. The simplest example of a field is perhaps the arithmetic of rational numbers. This has, however, an important additional property: its elements are either positive or negative. A field which has this property is called an *ordered field*. More precisely, an ordered field is a field certain of whose elements have a property known as being *positive*, satisfying the following laws:

(i) If a is not positive, then $-a$ is.
(ii) If a and $-a$ are both positive, then $a = 0$.
(iii) If a and b are positive, then so are $a + b$ and $a \cdot b$.

(This definition treats 0 as positive. See footnote, p. 15.)

Clearly the rational numbers form an ordered field under this definition. Ordered fields are investigated in Chapter **F**, and rational numbers are proved to form an ordered field in Chapter **G**.

The concepts of " greater " and " less " can be defined in terms of " positive ". Like " positive ", these words have two senses, an exclusive one and an inclusive one. a is *strictly less than* b (written as $a < b$) if $b - a$ is positive in the exclusive sense; a is *less than or equal to* b (written as $a \leqslant b$) if $b - a$ is positive in the inclusive sense.

Either sense is very easily defined in terms of the other. We can define the first in terms of the second by defining " $a < b$ " to mean " $a \leqslant b$ and $a \neq b$ "; we can define the second in terms of the first by defining $a \leqslant b$ to mean " $a < b$ or $a = b$ ". In the development, only one of these two is really necessary, the other being merely a convenience. Having chosen an inclusive definition of *positive*, we choose the inclusive *less than*.

An alternative treatment would have been to define an ordered field directly in terms of *less than*. An ordered field is then a field with a relation $\leqslant$ between its elements obeying the laws:

(*a*) If a and b are any elements of the field, then $a \leqslant b$ or $b \leqslant a$.
(*b*) If $a \leqslant b$ and $b \leqslant a$, then $a = b$.
(*c*) If $0 \leqslant a$ and $0 \leqslant b$, then $0 \leqslant a + b$ and $0 \leqslant a \cdot b$.

We could then define x to be positive if $0 \leqslant x$.

The reader may like, as an exercise, to deduce (i), (ii), and (iii) from this definition and (*a*), (*b*), and (*c*).

One more concept which can be defined in an ordered field is that of *absolute value* (or *numerical value* if the elements are numbers). The absolute value, $|x|$, of a positive element x is simply x; the absolute value of a negative element y is $-y$. This concept will be much used in the following chapters.

CHAPTER VII

Irrational Numbers

1. The field of rational numbers is a powerful tool. We can carry out all four operations of arithmetic in it, and all the laws of arithmetic are true for it. But arithmetic is not an end in itself, and when we try to apply it, we find that the rational numbers are not enough. We cannot, for example, solve the equation $x^2 = 2$; the ancient Greeks knew that there is no rational number whose square is 2. But such a number is required for geometry: if the sides of a square are of length 1, the length of its diagonal is a number whose square is 2. This, then, sets us a problem: to construct numbers which shall be solutions of such equations as $x^p = c$. Such numbers are called real numbers (not because they are any more real than the numbers we have met so far, but because they are in one sense more real than numbers we shall meet later). We can try the technique which we used before. Just as we constructed numbers like a/b to solve $b{\cdot}x = a$, so we can now construct numbers of the form $\sqrt[p]{c}$. But now we cannot satisfactorily define addition and multiplication: $\sqrt[3]{2} + \sqrt[2]{3}$ cannot really be defined to be anything except $\sqrt[3]{2} + \sqrt[2]{3}$. The best we can do in this way is to take as our system the set of all expressions which can be made up by applying the four operations of arithmetic to numbers of the form $\sqrt[p]{c}$. It is now very difficult to say when two numbers are equal, and practically impossible to say which of two is the greater. The whole system is cumbrous and complicated. And it is not good enough: we cannot in this way construct a number which will represent the length of the circumference of a circle of unit radius. The problem is in fact a difficult one, much deeper than those we solved in constructing the rational numbers. It was probably the discovery that there was no rational square root of 2 which accounted for the Greek neglect of arithmetic and concentration on geometry.

2. Fortunately there is an indirect way of solving the problem. In the field of rational numbers we can set up the familiar decimal system. In this system each rational number is represented by either a terminating or a recurring decimal. But it is easy to think of decimals which neither terminate nor recur: ·101001000100001 . . . , for example. It is natural to suppose that these represent numbers of some sort. And there is a strong hint that our missing roots are among them, for a method of finding square roots is described in textbooks of arithmetic. If we apply this method to a number which has no rational square-root we get a non-recurring non-terminating decimal from it. If we try to find the square root of 2, for example, we get 1·414 . . . and we can go on finding extra digits for as long as we like.

Now we could define our system to consist of all decimals. It would be tricky but not impossible to define addition and multiplication. We have, except for details, solved our problem. But this solution is an artificial one because it makes our number-system appear to depend in some fundamental way upon the number 10, the base of the decimal system. We can avoid this inelegance if we look a little more closely at a typical non-terminating decimal.

When we say that $\sqrt{2} = 1{\cdot}414\ldots$ we mean that there is a succession of rational numbers, namely 1, 1·4, 1·41, 1·414, . . . which get as close as we like to $\sqrt{2}$. 1 is within 1 of $\sqrt{2}$; 1·4 is within ·1 of $\sqrt{2}$; 1·41 is within ·01 of $\sqrt{2}$; and so on. Thus $\sqrt{2}$ is caught in a trap, and we can reduce the free play inside the trap to as little as we like by going far enough along the succession. In other words, if b is any positive non-zero number (no matter how small) we can find a number in the succession which is within b of $\sqrt{2}$.

This is, however, not quite enough. A succession like

$$1,\ 8,\ 1{\cdot}4,\ 9,\ 1{\cdot}41,\ 10,\ 1{\cdot}414,\ 11,\ \ldots$$

would have this property, but would not be much good as an approximation to $\sqrt{2}$. We therefore require a slightly stronger property: we want the succession not merely to *come* close to $\sqrt{2}$ but to *remain* close to $\sqrt{2}$. Let us say that the succession approximates to $\sqrt{2}$ *to within the standard* b (where b is positive and non-zero) if there is a number in the succession such that all later numbers in the succession are within b of $\sqrt{2}$. Then our succession will represent $\sqrt{2}$ if it is within all standards of approximation of $\sqrt{2}$; in other words, if, given any positive non-zero b, all the terms beyond a certain one are within b of $\sqrt{2}$. The argument still applies if any other number replaces $\sqrt{2}$.

The successions of rational numbers which represent a given number are those whose members come and remain arbitrarily close to the given number. And if their members are arbitrarily close to the same number, then they are arbitrarily close to each other. It turns out that the converse also is true: if the members of a succession are arbitrarily close to one another in this way (i.e. if, given any positive non-zero b, there is a member such that any two later members are within b of each other) then the succession represents a number. Such a succession is called a *Cauchy sequence.* Our way is now clear. We have only to state the conditions under which two Cauchy sequences are equivalent (i.e. represent the same number), and to define a real number as a set of equivalent Cauchy sequences of rational numbers.

3. The only properties of rational numbers which we require in framing our definitions are those which are common to all ordered fields. We therefore do not restrict our definitions to the ordered field of rational numbers but define a Cauchy number to be a set of equivalent Cauchy sequences of elements of an ordered field.

The set of Cauchy numbers can be proved to be an ordered field if we define *addition, multiplication,* and *positive* in the obvious ways. I shall call it a *Cauchy field.* It contains an ordered field isomorphic to the ordered field we started with. This consists in fact of the numbers which contain Cauchy sequences of the form

$$a, a, a, a, \ldots$$

(i.e. all the elements equal), which we make correspond to the element a of the original field. This situation is analogous to the one we have met several times already, and in the usual way we can now consider the original field to be embedded in its Cauchy field.

4. The Cauchy field of the field of rational numbers is the field of real numbers. The elements which correspond to rational numbers in the isomorphism mentioned above might be called *rational real numbers*, but except when we want to stress the fact that they are only isomorphic to, and not identical with, the rational numbers we started from, we shall abbreviate the phrase to *rational numbers.* The embedding of the rational numbers in the real numbers by this isomorphism leads to the idea of a *limit.* Let us denote by $\tilde{x}$ the rational real number corresponding to the rational number x. Suppose that one of the Cauchy sequences which make up the real number $\mathbf{a}$ is $a_1, a_2, a_3, \ldots.$ Then $\mathbf{a}$ is obviously related in some way to the succession $\tilde{a}_1, \tilde{a}_2, \tilde{a}_3, \ldots$ of rational real numbers. The relation is, in fact, that the succession approximates to $\mathbf{a}$ within every standard. When this is so we say that $\mathbf{a}$ is the *limit* of $\tilde{a}_1, \tilde{a}_2, \tilde{a}_3, \ldots.$ More generally, if $\mathbf{y}_1, \mathbf{y}_2, \mathbf{y}_3, \ldots$ is any succession of real numbers and if there is a real number $\mathbf{l}$ to which the succession approximates within every standard, then we say that $\mathbf{l}$ is the *limit* of the succession.

5. When we extended the arithmetic of whole numbers in order to be able to solve the equation $a + x = b$ for all whole numbers a and b, we found that we could solve the equation for *all* elements a and b in the new system, not merely for those which correspond to elements in the old system. If this had not been so we should have wanted to make a similar extension to the new system, and so *ad infinitum.* We have an analogous situation here: if we try to extend the Cauchy field by applying the same process again, the new field is isomorphic to the old. In other words, the Cauchy field contains the limits of all its Cauchy sequences. A field with this property is called *complete.*

6. An important idea connected with limits is that of continuity.

The intuitive idea of a continuous function f of one variable is simply that $f(\mathbf{x})$ should not suddenly change by large amounts when $\mathbf{x}$ changes by small amounts. For example, the function f defined by the following equations

$$\begin{cases} f(\mathbf{x}) = -\mathbf{g}\cdot\mathbf{x} & \text{if } \mathbf{0} \leqslant \mathbf{x} < \mathbf{1} \\ f(\mathbf{x}) = \mathbf{g}\cdot(\mathbf{2} - \mathbf{x}) & \text{if } \mathbf{1} \leqslant \mathbf{x} < \mathbf{3} \\ f(\mathbf{x}) = \mathbf{g}\cdot(\mathbf{4} - \mathbf{x}) & \text{if } \mathbf{3} \leqslant \mathbf{x} < \mathbf{5} \\ \text{and so on} & \end{cases}$$

is discontinuous. (The function arises in elementary dynamics. $f(\mathbf{x})$ is the upward velocity at time $\mathbf{x}$ of a ball dropped from a height $\frac{1}{2}\mathbf{g}$ on to a horizontal surface.)

When $\mathbf{x}$ is just less than $\mathbf{1}$, $f(\mathbf{x})$ is just greater than $-\mathbf{g}$, and can be made as near to $-\mathbf{g}$ as we like by taking $\mathbf{x}$ near enough to $\mathbf{1}$. When $\mathbf{x}$ is just greater than $\mathbf{1}$, $f(\mathbf{x})$ is just less than $+\mathbf{g}$, and can be made as near as we like to this by taking $\mathbf{x}$ near enough to $\mathbf{1}$. Therefore if $\mathbf{x}$ changes by a small amount from just less than $\mathbf{1}$ to just greater than $\mathbf{1}$—and we can make this amount as small as we like—$f(\mathbf{x})$ changes practically from $-\mathbf{g}$ to $+\mathbf{g}$. $f(\mathbf{x})$, in fact, makes a discontinuous change of amount $\mathbf{2g}$ where $\mathbf{x} = \mathbf{1}$.

A discontinuous function defined by a single formula is usually rather complicated, but a reader with the necessary technique will recognize that if we put $\phi(\mathbf{x})$ equal to the limit of the succession

$$\frac{\mathbf{x}-\mathbf{1}}{\mathbf{x}+\mathbf{1}}, \frac{\mathbf{x}^2-\mathbf{1}}{\mathbf{x}^2+\mathbf{1}}, \frac{\mathbf{x}^3-\mathbf{1}}{\mathbf{x}^3+\mathbf{1}}, \ldots$$

(i.e. define ϕ by the formula $\phi(\mathbf{x}) = \lim_{n\to\infty} (\mathbf{x}^n - \mathbf{1})/(\mathbf{x}^n + \mathbf{1})$) then ϕ is discontinuous where $\mathbf{x} = \mathbf{1}$.

A function f is *continuous* at $\mathbf{x} = \mathbf{a}$ if it does not suddenly change as $\mathbf{x}$ changes from value to value near $\mathbf{a}$. This means that if $\mathbf{a}_1$, $\mathbf{a}_2$, $\mathbf{a}_3$, . . . is any succession of values of $\mathbf{x}$ whose limit is $\mathbf{a}$, then the succession

$$f(\mathbf{a}_1), f(\mathbf{a}_2), f(\mathbf{a}_3), \ldots$$

must have $f(\mathbf{a})$ for its limit. We take this, then, as our definition of continuous function. An important property of a continuous function f which we can prove immediately is that if $f(\mathbf{x})$ is equal to $\mathbf{a}$ for some value of $\mathbf{x}$ and to $\mathbf{b}$ for some other value, then it must take all values between $\mathbf{a}$ and $\mathbf{b}$ as $\mathbf{x}$ varies from one value to the other. This property can be used to prove that, if n is a positive non-zero integer, every positive number $\mathbf{d}$ has an nth root; because we can show that $\mathbf{x}^n$ is a continuous function of $\mathbf{x}$, and can be made to range from $\mathbf{0}$ to a value greater than $\mathbf{d}$. Therefore for some value of $\mathbf{x}$ it must be equal to $\mathbf{d}$. This value is the nth root of $\mathbf{d}$.

7. The field of real numbers has one important property which not all Cauchy fields have; this is deduced from a property of rational numbers which not all ordered fields have, and is called *Archimedes' axiom*: given any element $\mathbf{a}$ of the field there is an integer n such that $n > \mathbf{a}$. (It occurs as lemma **G8** and theorem **J2**.)

CHAPTER VIII

Powers

1. The symbols $\mathbf{x}^2$, $\mathbf{x}^3$, $\mathbf{x}^4$, are used as abbreviations for $\mathbf{x}\cdot\mathbf{x}$, $(\mathbf{x}\cdot\mathbf{x})\cdot\mathbf{x}$, $((\mathbf{x}\cdot\mathbf{x})\cdot\mathbf{x})\cdot\mathbf{x}$, and so on, in elementary algebra, and obey the familiar index laws:

$$\mathbf{x}^m\cdot\mathbf{x}^n = \mathbf{x}^{m+n}$$
$$\mathbf{x}^m\cdot\mathbf{y}^m = (\mathbf{x}\cdot\mathbf{y})^m$$
$$(\mathbf{x}^m)^n = \mathbf{x}^{m\cdot n}$$

Our definition of $\mathbf{x}^n$ can be extended to cover negative numbers n by the usual methods. First, assume that there is a definition and that the index laws hold, and see what the definition must be. Second, state this definition and proceed to build up the theory from it.

If the index laws are to hold, we must have

$$\mathbf{x}^0\cdot\mathbf{x}^n = \mathbf{x}^{0+n} = \mathbf{x}^n. \text{ Therefore } \mathbf{x}^0 = 1$$

Then

$$\mathbf{x}^{-n}\cdot\mathbf{x}^n = \mathbf{x}^{-n+n} = \mathbf{x}^0 = \mathbf{1}. \text{ Therefore } \mathbf{x}^{-n} = \mathbf{1}/\mathbf{x}^n$$

Thus the first index law is enough to determine how we must define $\mathbf{x}^n$ for negative integers.

The definition is given (for positive and negative integers together) in Chapter **H**, and the index laws and other properties are deduced.

2. Now let us consider fractions. The third index law leads us to require

$$(\mathbf{x}^{p/q})^q = \mathbf{x}^{(p/q)\cdot q} = \mathbf{x}^p$$

Therefore $\mathbf{x}^{p/q}$ is the qth root of $\mathbf{x}^p$. This definition is made, and the corresponding theory developed, in Chapter **J**.

Finally we consider all real numbers. Any real number **u** can be considered as the limit of a Cauchy sequence $u_1, u_2, u_3, \ldots$ of rational numbers. And we can prove that the succession $\mathbf{x}^{u_1}, \mathbf{x}^{u_2}, \mathbf{x}^{u_3}, \ldots$ is a Cauchy sequence. It is natural to define $\mathbf{x}^{\mathbf{u}}$ to be the limit of this succession. Notice that we are not using our former technique; the three index laws, involving only products and sums, give us no way of getting from rational to irrational numbers. The reason for this is that the extension of the number-system from rational numbers to real numbers is different in kind from the previous extensions: we are embedding a field in a larger field; not embedding a "partial field" in a field.

CHAPTER IX

Complex Numbers

1. We have created numbers which will solve a large variety of equations, but by no means all. A simple equation with no solution in the field of real numbers is $\mathbf{x}^2 + \mathbf{1} = \mathbf{0}$.

A standard piece of elementary algebra is the solution of quadratic equations: sometimes they have two real solutions, sometimes not. We can prove as follows that if $\mathbf{x}^2 + \mathbf{1} = \mathbf{0}$ has a solution, then every quadratic equation has a solution. Let $\mathbf{i}$ be the solution of $\mathbf{x}^2 + \mathbf{1} = \mathbf{0}$; that is, suppose that $\mathbf{i}^2 + \mathbf{1} = \mathbf{0}$. Then if $\mathbf{a}\cdot\mathbf{x}^2 + \mathbf{b}\cdot\mathbf{x} + \mathbf{c} = \mathbf{0}$ is any quadratic equation, we can rewrite it as

$$\mathbf{x}^2 + \mathbf{2h}\cdot\mathbf{x} + \mathbf{g} = \mathbf{0}$$

where $\mathbf{g} = \mathbf{c}/\mathbf{a}$ and $\mathbf{h} = \frac{1}{2}\mathbf{b}/\mathbf{a}$. This can be rewritten as

(i) $$(\mathbf{x} + \mathbf{h})^2 + \mathbf{g} - \mathbf{h}^2 = \mathbf{0}$$

If $\mathbf{h}^2 - \mathbf{g}$ is positive, then it has a square root, and $-\mathbf{h} + \sqrt{(\mathbf{h}^2 - \mathbf{g})}$ is a solution of the equation (and so is the number $-\mathbf{h} - \sqrt{(\mathbf{h}^2 - \mathbf{g})}$ obtained by taking the other square root of $\mathbf{h}^2 - \mathbf{g}$). If $\mathbf{h}^2 - \mathbf{g}$ is not positive, then $\mathbf{g} - \mathbf{h}^2$ is, and therefore has a square root. Put $\mathbf{x} = -\mathbf{h} + \mathbf{i}\sqrt{(\mathbf{g} - \mathbf{h}^2)}$. This is a solution of (i) because, if we substitute, the left-hand side becomes

$$\begin{aligned} &(\mathbf{i}\cdot\sqrt{(\mathbf{g} - \mathbf{h}^2)})^2 + \mathbf{g} - \mathbf{h}^2 \\ &= \mathbf{i}^2\cdot(\mathbf{g} - \mathbf{h}^2) + \mathbf{g} - \mathbf{h}^2 \\ &= (\mathbf{g} - \mathbf{h}^2)\cdot(\mathbf{i}^2 + \mathbf{1}) \\ &= (\mathbf{g} - \mathbf{h}^2)\cdot\mathbf{0} \\ &= \mathbf{0} \end{aligned}$$

What is more surprising (and more difficult to prove) is that if our field contains the real numbers and a number $\mathbf{i}$ such that $\mathbf{i}^2 + \mathbf{1} = \mathbf{0}$, then it contains a solution of any equation of the form

$$\mathbf{a} + \mathbf{b}\cdot\mathbf{x} + \mathbf{c}\cdot\mathbf{x}^2 + \mathbf{d}\cdot\mathbf{x}^3 + \ldots + \mathbf{k}\cdot\mathbf{x}^n = \mathbf{0}$$

2. Let us adopt the usual technique of assuming that a number such as $\mathbf{i}$ exists, and seeing what properties our field must have. A field must contain the sum and product of any two elements. Therefore if it contains $\mathbf{i}$, and if $\mathbf{x}$ and $\mathbf{y}$ are any real numbers, it must contain $\mathbf{x} + \mathbf{i}\cdot\mathbf{y}$.

On the other hand, the set of all numbers of the form $\mathbf{x} + \mathbf{i}\cdot\mathbf{y}$ where $\mathbf{x}$ and $\mathbf{y}$ are real numbers, $\mathbf{i}^2 = -\mathbf{1}$, and the laws of arithmetic are obeyed, does contain the sum and product of all its elements:

(ii)
$$\begin{aligned}(\mathbf{a} + \mathbf{i}\cdot\mathbf{b}) + (\mathbf{c} + \mathbf{i}\cdot\mathbf{d}) &= (\mathbf{a} + \mathbf{c}) + (\mathbf{i}\cdot\mathbf{b} + \mathbf{i}\cdot\mathbf{d}) \\ &= (\mathbf{a} + \mathbf{c}) + \mathbf{i}\cdot(\mathbf{b} + \mathbf{d})\end{aligned}$$

$\mathbf{a} + \mathbf{c}$ and $\mathbf{b} + \mathbf{d}$ are real numbers, and so $(\mathbf{a} + \mathbf{c}) + \mathbf{i}\cdot(\mathbf{b} + \mathbf{d})$ is a number of the form $\mathbf{x} + \mathbf{i}\cdot\mathbf{y}$.

(iii)
$$\begin{aligned}(\mathbf{a} + \mathbf{i}\cdot\mathbf{b})\cdot(\mathbf{c} + \mathbf{i}\cdot\mathbf{d}) &= \mathbf{a}\cdot\mathbf{c} + \mathbf{i}\cdot(\mathbf{a}\cdot\mathbf{d} + \mathbf{b}\cdot\mathbf{c}) + \mathbf{i}^2\cdot\mathbf{b}\cdot\mathbf{d} \\ &= (\mathbf{a}\cdot\mathbf{c} - \mathbf{b}\cdot\mathbf{d}) + \mathbf{i}\cdot(\mathbf{a}\cdot\mathbf{d} + \mathbf{b}\cdot\mathbf{c})\end{aligned}$$

which again is of the form $\mathbf{x} + \mathbf{i}\cdot\mathbf{y}$.

We can go further, and prove that the set of all numbers of the form $\mathbf{x} + \mathbf{i}\cdot\mathbf{y}$ is a field. Therefore, if we can find a valid definition of $\mathbf{i}$, the set will be the field that we require.

3. The fact that the required numbers are of the form $\mathbf{x} + \mathbf{i}\cdot\mathbf{y}$—an expression involving two real numbers—suggests that, as before, we should consider sets of ordered pairs. Under what conditions should we have $\mathbf{a} + \mathbf{i}\cdot\mathbf{b} = \mathbf{c} + \mathbf{i}\cdot\mathbf{d}$?

If $\mathbf{a} + \mathbf{i}\cdot\mathbf{b} = \mathbf{c} + \mathbf{i}\cdot\mathbf{d}$, then $(\mathbf{a} - \mathbf{c}) = \mathbf{i}\cdot(\mathbf{d} - \mathbf{b})$ and so

$$\begin{aligned}(\mathbf{a} - \mathbf{c})^2 &= \mathbf{i}^2\cdot(\mathbf{d} - \mathbf{b})^2 \\ &= -(\mathbf{d} - \mathbf{b})^2\end{aligned}$$

We have a positive quantity equal to a negative quantity, and so both must be zero. Then $\mathbf{a} = \mathbf{c}$ and $\mathbf{b} = \mathbf{d}$. Therefore $\mathbf{a} + \mathbf{i}\cdot\mathbf{b}$ and $\mathbf{c} + \mathbf{i}\cdot\mathbf{d}$ are equal if and only if they are the same. This means that we need not consider *sets* of ordered pairs of real numbers; the ordered pairs themselves will do. We define a complex number, then, to be an ordered pair of real numbers. Equations (ii) and (iii) show us how we must define addition and multiplication:

$$\begin{aligned}(\mathbf{a}, \mathbf{b}) + (\mathbf{c}, \mathbf{d}) &= (\mathbf{a} + \mathbf{c}, \mathbf{b} + \mathbf{d}) \\ (\mathbf{a}, \mathbf{b})\cdot(\mathbf{c}, \mathbf{d}) &= (\mathbf{a}\cdot\mathbf{c} - \mathbf{b}\cdot\mathbf{d}, \mathbf{a}\cdot\mathbf{d} + \mathbf{b}\cdot\mathbf{c})\end{aligned}$$

From these definitions we can now prove that the set of ordered pairs is a field, by verifying the axioms. To verify, for example, the associative law of addition we see that

$$((\mathbf{a}, \mathbf{b}) + (\mathbf{c}, \mathbf{d})) + (\mathbf{e}, \mathbf{f}) = ((\mathbf{a} + \mathbf{c}, \mathbf{b} + \mathbf{d})) + (\mathbf{e}, \mathbf{f})$$

by definition of addition

$$= ((\mathbf{a} + \mathbf{c}) + \mathbf{e}, (\mathbf{b} + \mathbf{d}) + \mathbf{f})$$

by definition of addition

$$= (\mathbf{a} + (\mathbf{c} + \mathbf{e}), \mathbf{b} + (\mathbf{d} + \mathbf{f}))$$

by the associative law of addition for real numbers

$$= (\mathbf{a}, \mathbf{b}) + (\mathbf{c} + \mathbf{e}, \mathbf{d} + \mathbf{f})$$
$$= (\mathbf{a}, \mathbf{b}) + ((\mathbf{c}, \mathbf{d}) + (\mathbf{e}, \mathbf{f}))$$

Some of the other verifications are long, but they are all straightforward.

As usual, our new field contains a field isomorphic to the field with which we started, namely the set of complex numbers of the form $(\mathbf{x}, \mathbf{0})$. These may be called *real complex numbers.*

Finally, our complex field has the usual completeness property. Not only can we solve the equation $\mathbf{a} + \ldots + \mathbf{k}\cdot\mathbf{x}^n = \mathbf{0}$ when $\mathbf{a}, \ldots, \mathbf{k}$ are real numbers; we can also solve it when they are complex numbers. Therefore a second extension along the same lines would give us nothing new.

The field of complex numbers is not an ordered field—in any ordered field $\mathbf{x}^2 + \mathbf{1} = \mathbf{0}$ is unsolvable (because $\mathbf{x}^2$ is positive, whereas $-\mathbf{1}$ is strictly negative, and so we cannot have $\mathbf{x}^2 = -\mathbf{1}$).

4. The extension to the complex numbers brings the present development of the number-system to an end. The reader will probably wonder whether this is the last extension which can be made. It is not. There is, for example, the quaternion system, which is most simply described as the set of elements of the form $\mathbf{x} + \mathbf{i}\cdot\mathbf{y} + \mathbf{j}\cdot\mathbf{z} + \mathbf{k}\cdot\mathbf{w}$ where $\mathbf{x}$, $\mathbf{y}$, $\mathbf{z}$, and $\mathbf{w}$ are real numbers, and $\mathbf{i}$, $\mathbf{j}$, and $\mathbf{k}$ are analogous to the $\mathbf{i}$ of the complex number field. (In fact, $\mathbf{i}^2 = \mathbf{j}^2 = \mathbf{k}^2 = -\mathbf{1}$, $\mathbf{i}\cdot\mathbf{j} = -\mathbf{j}\cdot\mathbf{i} = \mathbf{k}$, $\mathbf{j}\cdot\mathbf{k} = -\mathbf{k}\cdot\mathbf{j} = \mathbf{i}$, $\mathbf{k}\cdot\mathbf{i} = -\mathbf{i}\cdot\mathbf{k} = \mathbf{j}$.) We paid a price for the extension to complex numbers: the price was the order-relation. We pay another price for the extension to quaternions: the commutative law of multiplication must be given up, though all the other field-axioms hold.

There is a sense in which the various extensions of the number-system reach their peak in the ordered field of real numbers. After that we have to make concessions if we are to extend the system any further. There is probably a connection between this fact and the fact that the real number-system is the last system which appeals to common sense. No one doubts the existence and usefulness of negative, fractional, and irrational numbers; but, though the usefulness of complex numbers is unquestioned, it is only too easy for the non-mathematician to doubt their existence: it is " obvious " that $-\mathbf{1}$ has no square root. A strict development, constructing the complex numbers as pairs of real numbers, makes their existence logically certain, but even so there is still a context in which they seem less " real " than real numbers. This context is coördinate geometry.

In elementary geometry, a Cartesian coördinate system can be set up in which the coördinates of a point are two real numbers. By algebraic methods we can solve such geometrical problems as finding the intersection of a straight line and a circle.

For example, the line $\mathbf{x} = \mathbf{a}$ cuts the circle $\mathbf{x}^2 + \mathbf{y}^2 = \mathbf{r}^2$ in the points whose coördinates are the solutions of these two equations, namely $(\mathbf{a}, \sqrt{(\mathbf{r}^2 - \mathbf{a}^2)})$. The points have real coördinates if and only if $\mathbf{a} \leqslant \mathbf{r}$, i.e. if and only if the line does cut the circle. If the line does not cut the circle, their points of intersection do not exist—they are imaginary. In this way we have an interpretation of complex numbers as coördinates of imaginary points. For this reason non-real complex numbers were once called " imaginary " or " impossible ". These words are no longer used, but there is a relic of them in the description of a number $\mathbf{x} + \mathbf{i}\cdot\mathbf{y}$ for which $\mathbf{x} = \mathbf{0}$ as a " pure imaginary ".

The fact that, in one interpretation, complex numbers correspond to imaginary objects cannot, of course, make the complex numbers themselves any less real (using the word " real " here in its everyday sense, not in the technical sense). $\sqrt{(-\mathbf{1})}$ is just as real as $\mathbf{1}$ itself. The reality of the number $\mathbf{1}$ is, however, not a mathematical problem. It must be left to the philosophers.

CHAPTER X

Verification of the Axioms

1. In Chapter I we discussed the use of numbers for counting, and in Chapter II we showed how the common-sense properties of counting led us to postulate three axioms. On these axioms we based the whole of arithmetic: the details are given in Chapters **A**–**K**, and the reasons for the definitions adopted are explained in Chapters II–IX. Now that we have the complete system before us, let us consider more closely what is meant by a number, and verify that numbers do indeed satisfy the axioms.

It may be as well to remark that arithmetic can stand on its own as an abstract mathematical structure; but without some such consideration as that proposed there will be no connection between abstract arithmetic and arithmetic as used by a grocer or a physicist. This is what E. Landau (in *Grundlagen der Analysis*) means when he says that " one " is simply a word of the English language (or rather, that " ein " is simply a word of the German language): he refrains from forging the link between the " one " of common-sense arithmetic and the 1 which forms part of his abstract system.

There is a more subtle disadvantage to the purely abstract approach. Suppose that we took the symbol [. . .] for the abstract 1, and defined the successor of any element to be an element with one more dot. Then we should get a perfectly self-consistent arithmetic, whose elements are [. . .], [. . . .], [. . . . .], etc., and in which addition and multiplication are commutative, associative, etc., but in which [. . .] + [. . .] is equal to [. . . .] and not to [. . . . . .]. This arithmetic is isomorphic to the familiar arithmetic of whole numbers, but by the misleading matching

$$1 \longleftrightarrow [\,.\,.\,.\,]$$
$$2 \longleftrightarrow [\,.\,.\,.\,.\,]$$
etc.

instead of by the natural matching

$$1 \longleftrightarrow [\,.\,]$$
$$2 \longleftrightarrow [\,.\,.\,]$$
etc.

In fact, by the very nature of the axioms, anything isomorphic to the arithmetic of whole numbers is bound to satisfy them, and if we want our system to be practically recognizable as (not merely abstractly equivalent to) the familiar arithmetic, then we need something besides the purely abstract axiomatic development. The purely abstract methods can prove statements like $(1 + 2) + 1 = 1 + (2 + 1)$, but not statements like " If you and I are here, then two persons are here "; the wider methods can prove both sorts of statement.

2. What do we mean by a number? We have seen that a set contains three objects if and only if it is matchable with the dots in [. . .], i.e. with the little black spots which you are actually looking at now. In other words, the number *three* is something common to all sets matchable with this one. We might, then, define a number to be something common to all sets matchable with a given set. We might, slightly daring, define the number of members of a set S as " that which is common to all sets matchable with S ". A thoughtful reader, however, will want to examine this phrase more closely. Just *what* is common to all these sets? Various answers have been given. For example, we could take the *set of all sets matchable with the given set.* But the words " set of all sets . . ." are dangerous, as is shown by the famous Russell paradox,* and before we can use them we must be sure that the logic which underlies our thought is sound. In fact, we must investigate elementary logic in the way that the present book investigates elementary arithmetic—and logic presents a far deeper problem. One investigation is contained in the *Principia Mathematica* of Russell and Whitehead and explained in Russell's *Introduction to Mathematical Philosophy.*

Let us suppose, however, that we are content to leave this problem to the logicians, and to assume that there is some property (even though we do not know precisely what) common to a set of matchable sets. This gives us a definition of number in general. We now want to define specific numbers: the number 1, for example. It could be defined as

* The set of all teaspoons does not contain itself as a member, because it is not a teaspoon. And this is the normal situation. But certain exceptional sets may contain themselves as members: the set of all objects of thought, for example, or (rather obviously) the set of all sets. We ask: what about *the set of all sets which are not members of themselves*? Is it a member of itself or not?

Suppose first that it is a member of itself. Because its members are *sets which are not members of themselves,* this means that it is not a member of itself, contradicting the supposition, which must therefore be wrong. The set, then, is not a member of itself.

We are now in a difficulty, because the fact that it is not a member of itself, i.e. is not one of those sets which are not members of themselves, implies that it is a member of itself. And this we have just shown to be untrue.

the number of dots in [.]; in fact, it was so defined in Chapter I. This is rather like defining a yard in terms of a certain metal bar kept in a certain safe place. It is quite practicable, but does not seem satisfactory for such an abstract concept as a number. We want a definition in terms of pure logic, not in terms of certain marks on a certain piece of paper.

3. What do we mean when we say that a set contains just one member, or that there is just one element with a given property? We say, for example, that there is just one number x with the property $2 \cdot x = 3$. How do we prove this? We show first that the number $\frac{3}{2}$ has the desired property, by proving that $2 \cdot \frac{3}{2} = 3$. This proves that there is *at least one* number with the property. Then we show that if $2 \cdot x = 3$, it follows that $x = \frac{3}{2}$, so that $\frac{3}{2}$ is the only solution of the equation. This shows that there is *at most one* number with the property. In a similar way, we can make a formal definition of the number 1 as follows:

Definition: A set S is said to be a *unit set* if it has a member and if it is impossible for both x and y to be members of S unless $x = y$. The number 1 is the number of members of a unit set.

In a formal development we should, of course, have to show that all unit sets are matchable with each other before we wrote the second sentence.

4. Similarly, we can define 2 as the number of members of any set T which has the following properties: T has members x and y such that $x \neq y$; and if z belongs to T, then either $z = x$ or $z = y$. And so on. In this way we can define any specific number, though the definition of *one million* along these lines would be cumbrous. But our immediate aim is only to show that common-sense numbers satisfy our formal axioms, and to do this we need only define 0 and *successor*.

Definition: The set which has no member is the *null* set.
Definition: The number of members of the null set is 0.
Definition: The number m is a *successor* of the number n if it is the number of members of a set T satisfying the following conditions: there is a set S with n members, and an element x which belongs to T but not to S; every member of S belongs to T; and if z belongs to T, then either $z = x$ or z belongs to S.

We shall now prove that any such T is matchable with any other such set T^*. Suppose that T is as above, and that T^* contains x^*

and the members of a set S^* with n members, and so on, as above. Because S and S^* have the same number of members, there is a matching between them. We can define a matching between T and T^* by letting each member of S correspond to its mate in the matching between S and S^*, and x correspond to x^*.

It now follows that if m and m^* are successors of n, then $m = m^*$, because they are the numbers of members of matchable sets. In other words, n has just *one* successor. In fact, m is not merely *a* successor, but *the* successor, of n.

If we now define the set of all whole numbers to be the least set which contains 0 and which contains the successor of every member, then it is not hard to prove that axioms **A1a**, **b**, and **c** are satisfied.

5. To complete the connection between these numbers and the number-system in the formal development, we have to show that addition and multiplication agree with common sense: that is, that a set with m members and a set with n members (all different from the members of the first set) have between them $m + n$ members, and that m sets with n members each (all different) have between them $m \cdot n$ members. To do this we first prove that if S is matchable with S^* and T with T^*, and if S and T have no common member, and S^* and T^* have no common member, then the set which consists of the members of S together with those of T is matchable with the set which consists of the members of S^* together with those of T^*. We can now define the " sum " $m \oplus n$ of m and n to be the number of members of a set which is made up of a set of m members and a set of n members with no member in common. (No matter which set of m members and which set of n (different) members are chosen, the sets we get by putting them together are all matchable and so define the same number $m \oplus n$.) It follows easily that $m \oplus 0 = m$ and, if we denote the successor of x by x', that $m \oplus n' = (m \oplus n)'$. Then, by note **A3a** (which says that addition is uniquely defined by these two equations), we see that $\oplus$-addition is the same as $+$-addition. A similar argument holds for multiplication.

6. We must be careful about one point. Our definition of *number* makes considerable use of matchings, and a natural definition of *matching* would run something like this: it is a correspondence in which each member of a given set corresponds to exactly one member of another set. This will not do. We have defined *one* in terms of *matching*; we cannot also define *matching* in terms of *one*. But it is not hard to give an independent definition:

A correspondence $x \longleftrightarrow x^*$ is a *matching* if, whenever $x = y$, it follows that $x^* = y^*$, and, whenever $x^* = y^*$, it follows that $x = y$.

A *function* is rather similar to a matching. A function f of one variable is in fact a correspondence $x \to f(x)$ in which if $x = y$ then $f(x) = f(y)$. (Compared with a matching, it simply lacks the condition that if $f(x) = f(y)$ then $x = y$.) Some textbooks admit multi-valued functions: notice that this definition excludes them. (The two-valued square-root function is excluded, for example, because $4 = 4$ but $2 \neq -2$, although $2 = \sqrt{4}$ and $-2 = \sqrt{4}$. Multi-valued functions, in fact, play havoc with the logical axiom that things equal to the same thing are equal to one another. $2 = \sqrt{4} = -2$!) In this book, *function* will always mean *single-valued function.* Two functions f and g are said to be *equal* if $f(x)$ is equal to $g(x)$ for every x for which either is defined.

An operation θ is similarly a correspondence

$$(x, y) \to \theta(x, y)$$

(between ordered pairs of elements and elements) for which $x = x^*$ and $y = y^*$ implies $\theta(x, y) = \theta(x^*, y^*)$. Two operations θ and ϕ are said to be *equal* if $\theta(x, y) = \phi(x, y)$ for all ordered pairs (x, y) for which either $\theta(x, y)$ or $\phi(x, y)$ is defined.

A function of two variables is the same thing as an operation (unless multi-valued functions are allowed—operations are never multi-valued); the difference is only that " operation " is commonly used in algebra, " function " in analysis.

CHAPTER XI

Alternative Treatments

1. The present treatment is by no means the only possible one. A few of the more obvious alternatives ought to be mentioned.

First, the considerations we have taken last (in Chapter X) could be taken first: numbers could have been defined in terms of sets. The axioms of Chapter **A** (and Chapter II) would then have appeared as theorems. Some of the proofs would remain the same, but others (for example, the commutative law of addition) could be proved directly.

2. Next, the new types of number could have been introduced in a different order. If we had applied the embedding theory of Chapter **B** to the multiplicative hemigroup of non-zero whole numbers we should have arrived at the positive non-zero rational numbers. An additive application of the embedding theory to this would then give us the complete rational number-system. Or we could develop the positive real numbers from the positive rational numbers and introduce negative numbers last of all. This is the order which E. Landau adopts in his *Grundlagen der Analysis*; it corresponds more closely to the historical order, for negative numbers were not thought of until long after fractions were familiar mathematical elements. The reasons for adopting the non-historical order are: (i) to have available a different treatment from Landau's classical one, (ii) to separate the algebraic technique (introduction of fractions and negative numbers) from the analytical technique (the use of Cauchy sequences or some similar infinite process), as far as possible, and (iii) to define the system of integers, a system which is one of some importance in higher mathematics (in the Theory of Numbers) and which does not appear in the other treatment. Further, the introduction of the fractions then furnishes a good example of the algebraic process of embedding an integral domain in a field. (An integral domain is an important algebraic system: it is an additive commutative group with a second operation written as multiplication, obeying the commutative, associative, and distributive laws, and the law that if $x \cdot y = 0$ then either x or y is 0). In his *Introduction to Abstract*

Algebra, C. C. MacDuffee exploits this treatment by developing the system of integers in some detail.

3. There are many ways of introducing the real numbers once the rationals have been constructed. Perhaps the best is the use of Dedekind sections. This method is described in G. H. Hardy's *Pure Mathematics* and E. Landau's *Grundlagen der Analysis*, and the existence of these two well-known treatments was one of the reasons for not using the method here. Another way—hinted at in Chapter VII—is by the use of infinite decimals, and this (or rather a similar treatment using radix fractions to the base 2 instead of 10) is probably the best really elementary method.

The process of embedding could also have been carried out differently. We constructed the system N of whole numbers and from it deduced the system I of integers, containing a subset N' (namely the positive integers) isomorphic to N. We then proceeded to ignore N and to let the positive integers play the part of the whole numbers. The alternative is to include N in a new system, which we could do as follows. Let J consist of the elements of N together with those elements of I which are not in N'. If n is an element of N, let n' be the element of N' which corresponds to it in the isomorphism between the two. We define operations and relations in J as follows:

$$x + y \text{ is as in } I \text{ if } x \in I \text{ and } y \in I$$
$$x + y \text{ is } x' + y \text{ if } x \in N \text{ and } y \in I$$
$$x + y \text{ is } x' + y' \text{ if } x \in N \text{ and } y \in N$$

provided that $x + y$, $x' + y$, $x' + y'$ are in J. If any of them is not in J, then it must be in N', and we take in its place the corresponding element of N.

Multiplication is defined similarly:

If $x \in I$, then x is positive in J if and only if it is positive in I.

If $x \in N$, then x is positive in J if and only if x' is positive in J.

4. We could have adopted a completely different way of developing the whole numbers from Peano's axioms. We chose to deal with addition and multiplication first. If we had then wanted to deal with the order-relation, we could have done so by defining "$x \geqslant y$" to mean "there is a z such that $x = y + z$". The alternative is to deal with the order-relation first. This treatment will be found in R. Dedekind's *Was sind und was sollen die Zahlen?*

5. All these treatments start from Peano's axioms for the whole numbers—and no better axioms for the whole numbers have ever been devised. But instead of starting there and constructing the integers, rationals, and so on, step by step, we could have started with axioms for the integers, and built up our system from them (this course is adopted by Birkhoff and MacLane in *A Survey of Modern Algebra*); or we could have started from the rationals—a good common-sense starting point, which is implicitly used by G. H. Hardy, who, in *Pure Mathematics*, assumes a knowledge of the properties of the rational numbers, although he does not explicitly state any axioms. We could even give an axiom-system for the real numbers (it would be simply the axiom-system for a complete Archimedean ordered field) leaving only the complex numbers to be constructed.

6. Finally, if we had had sufficient algebraic technique at our disposal, we could have defined the complex numbers to be the elements of the root field of the equation $x^2 + 1 = 0$ over the rational field. This method is described in A. A. Albert's *Modern Higher Algebra*.

Part II

SYSTEMATIC TREATMENT

CHAPTER **A**

Whole Numbers

1. Definitions (of *succession*, *successor*, and *whole number*): A *succession* is a set N of elements such that for each x of N there is a unique element of N, which we denote by x', with the following properties:

a. If $x' = y'$ then $x = y$.

b. There is an element of N, which we denote by 0, such that, for every x of N, $x' \neq 0$.

c. If M is a set which contains 0 and which contains x' whenever it contains x, then M contains every element of N.

x' is the *successor* of x.

A succession in which addition and multiplication are suitably defined (i.e. have the properties described in **3** and **9**, below) is a succession of *whole numbers*.

Notation: In Chapters **A**, **C**, **E**, and **G**, small italic letters denote whole numbers.

2. Theorem: *If* $x \neq 0$, *then there is a* y *such that* $x = y'$.

Proof: Let M consist of 0 and of all x for which there is such a y. Then $0 \in M$. If $x \in M$, then $x' \in M$, because $x' = x'$. Therefore, by **1c**, M contains every whole number, and so the theorem is true.

Note: **1b** states that 0 is a non-successor; **2**, that it is the *only* non-successor.

3. Theorem: *There is just one operation* θ *such that, for every* x *and* y,

(i) $$\theta(x, 0) = x$$

and

(ii) $$\theta(x, y') = \theta(x, y)'.$$

Proof: If θ exists, let ϕ be an operation such that

(iii) $$\phi(x, 0) = x \quad \text{for every } x$$

and

(iv) $$\phi(x, y') = \phi(x, y)' \quad \text{for every } x \text{ and } y.$$

Let M be the set of y for which $\phi(x, y) = \theta(x, y)$ for every x. Then $0 \in M$, because $\phi(x, 0) = x = \theta(x, 0)$ for every x, by (iii) and (i).

If $y \in M$, then
$$\begin{aligned} \phi(x, y') &= \phi(x, y)' && \text{by (iv)} \\ &= \theta(x, y)' && \text{because } y \in M \\ &= \theta(x, y') && \text{by (ii).} \end{aligned}$$

Therefore $y' \in M$. Therefore, by **1c**, M contains all whole numbers, and so $\phi(x, y) = \theta(x, y)$ for every x and y; that is, there is at *most* one operation satisfying (i) and (ii). It remains to prove that there is at *least* one.

Let M be the set of x for which there is, for each y, a whole number $\theta(x, y)$ such that (i) and (ii) are true. If we let

(v) $$\theta(0, y) = y \qquad \text{for every } y,$$

then $$\theta(0, 0) = 0 \qquad \text{by (v).}$$

Therefore (i) is true when $x = 0$. And

$$\begin{aligned} \theta(0, y') &= y' && \text{by (v)} \\ &= \theta(0, y)' && \text{by (v).} \end{aligned}$$

Therefore (ii) is true when $x = 0$. Therefore $0 \in M$.

If $z \in M$, let

(vi) $$\theta(z', y) = \theta(z, y)' \qquad \text{for every } y.$$

($\theta(z, y)$ is defined, because $z \in M$.) Then

$$\begin{aligned} \theta(z', 0) &= \theta(z, 0)' && \text{by (vi)} \\ &= z' && \text{by (i)} \end{aligned}$$

because $z \in M$. Therefore (i) is true when $x = z'$. And

$$\begin{aligned} \theta(z', y') &= \theta(z, y')' && \text{by (vi)} \\ &= \theta(z, y)'' && \text{by (ii), because } z \in M \\ &= \theta(z', y)' && \text{by (vi).} \end{aligned}$$

Therefore (ii) is true when $x = z'$, and so $z' \in M$.

Therefore, by **1c**, M contains every whole number, and so (i) and (ii) are true for every x and y.

We may now talk about *the* operation for which (i) and (ii) are true; (v) and (vi) are also true for it.

Definition (of *sum* of whole numbers): The number $\theta(x, y)$ just defined is the *sum* of x and y, and will be written as $x + y$. (i), (ii), (v), and (vi) become

a. $$x + 0 = x = 0 + x$$

and

b. $$x + y' = (x + y)' = x' + y.$$

4. Theorem: $(x + y) + z = x + (y + z)$ *for every* x, y, *and* z.

Proof: Let M be the set of z for which $(x + y) + z = x + (y + z)$ for every x and y.

$$\begin{aligned} (x + y) + 0 &= x + y && \text{by } \mathbf{3a} \\ &= x + (y + 0) && \text{by } \mathbf{3a}. \end{aligned}$$

Therefore $0 \in M$.

If $z \in M$, then
$$\begin{aligned} (x + y) + z' &= ((x + y) + z)' && \text{by } \mathbf{3b} \\ &= (x + (y + z))' && \text{because } z \in M \\ &= x + (y + z)' && \text{by } \mathbf{3b} \\ &= x + (y + z') && \text{by } \mathbf{3b}. \end{aligned}$$

Therefore $z' \in M$; and so, by **1c**, the theorem is true.

5. Theorem: $x + y = y + x$ *for every* x *and* y.

Proof: Let M be the set of x for which $x + y = y + x$ for every y.

$$0 + y = y + 0 \qquad \text{by } \mathbf{3a}.$$

Therefore $0 \in M$.

If $x \in M$, then

$$\begin{aligned} x' + y &= (x + y)' && \text{by } \mathbf{3b} \\ &= (y + x)' && \text{because } x \in M \\ &= y + x' && \text{by } \mathbf{3b}. \end{aligned}$$

Therefore $x' \in M$; and so, by **1c**, the theorem is true.

6. Theorem: *If* $y + x = z + x$, *then* $y = z$.

Proof: Let M be the set of x for which this is true for every y and z. $0 \in M$, by **3a**.

If $x \in M$, and if there is a y and a z such that

$$y + x' = z + x'$$

then $(y + x)' = (z + x)'$ by **3b**.

Therefore $y + x = z + x$ by **1a**.

Therefore $y = z$ because $x \in M$.

Therefore $x' \in M$; and so, by **1c**, the theorem is true.

7. Theorem: *If* $x + y = 0$, *then* $x = y = 0$.

Proof: If $y \neq 0$, then there is a u such that $y = u'$, by **2**. Then

$$x + y = x + u' = (x + u)' \quad \text{by } \mathbf{3b}$$
$$\neq 0 \quad \text{by } \mathbf{1b}.$$

Therefore y cannot *not* be 0. Then $x + 0 = 0$, and so $x = 0$, by **3a**.

8. Theorem: *If x and y are any whole numbers, then either $x = u + y$ for some u, or $y = v + x$ for some v.*

Proof: Let M be the set of x for which this is true for every y. Then $0 \in M$, because, by **3a**, $y = y + 0$, and we take y for the v. If $x \in M$, then either (i) $x = u + y$ or (ii) $y = v + x$ and $v \neq 0$ or (iii) $y = v + x$ and $v = 0$.

In case (i), $x' = (u + y)' = u' + y$, by **3b**.

In case (ii), $v = u'$ for some u, by **2**. Then $y = u' + x = u + x'$, by **3b**.

In case (iii), $y = 0 + x = x$, by **3a**. Then $0' + y = 0' + x = x'$, by **3a** and **3b**.

In each case, then, either $x' = u + y$ or $y = v + x'$ for some u or v. Therefore $x' \in M$; and so, by **1c**, the theorem is true.

9. Theorem: *There is just one operation θ such that, for every x and y,*

(i) $$\theta(x, 0) = 0$$

and

(ii) $$\theta(x, y') = \theta(x, y) + x.$$

Proof: If θ exists, let ϕ be an operation such that

(iii) $$\phi(x, 0) = 0 \quad \text{for every } x$$

and

(iv) $$\phi(x, y') = \phi(x, y) + x \quad \text{for every } x \text{ and } y.$$

Let M be the set of y for which $\phi(x, y) = \theta(x, y)$ for every x.

$$
\begin{aligned}
0 \in M, \text{ because} \quad \phi(x, 0) &= 0 && \text{by (iii)} \\
&= \theta(x, 0) && \text{by (i).} \\
\text{If } y \in M, \text{ then} \quad \phi(x, y') &= \phi(x, y) + x && \text{by (iv)} \\
&= \theta(x, y) + x && \text{because } y \in M \\
&= \theta(x, y') && \text{by (ii).}
\end{aligned}
$$

Therefore $y' \in M$. Therefore, by **1c**, $\phi(x, y) = \theta(x, y)$ for every x and y; that is, there is at *most* one operation satisfying (i) and (ii). It remains to prove that there is at *least* one.

Let M be the set of x for which there is, for each y, a whole number $\theta(x, y)$ such that (i) and (ii) are true.

If we let

(v) $$\theta(0, y) = 0 \quad \text{for every } y$$

then $$\theta(0, 0) = 0.$$

Therefore (i) is true when $x = 0$. And

$$
\begin{aligned}
\theta(0, y') &= 0 && \text{by (v)} \\
&= \theta(0, y) && \text{by (v)} \\
&= \theta(0, y) + 0 && \text{by } \mathbf{3a}.
\end{aligned}
$$

Therefore (ii) is true when $x = 0$. Therefore $0 \in M$.

If $z \in M$, let

(vi) $$\theta(z', y) = \theta(z, y) + y \quad \text{for every } y.$$

($\theta(z, y)$ is defined because $z \in M$.) Then

$$
\begin{aligned}
\theta(z', 0) &= \theta(z, 0) + 0 && \text{by (vi)} \\
&= 0 + 0 && \text{by (i), because } z \in M \\
&= 0 && \text{by } \mathbf{3a}.
\end{aligned}
$$

Therefore (i) is true when $x = z'$. And

$$
\begin{aligned}
\theta(z', y') &= \theta(z, y') + y' && \text{by (vi)} \\
&= (\theta(z, y) + z) + y' && \text{by (ii), because } z \in M \\
&= \theta(z, y) + (z + y') && \text{by } \mathbf{4} \\
&= \theta(z, y) + (y' + z) && \text{by } \mathbf{5} \\
&= \theta(z, y) + (y + z') && \text{by } \mathbf{3b} \\
&= (\theta(z, y) + y) + z' && \text{by } \mathbf{4} \\
&= \theta(z', y) + z' && \text{by (vi).}
\end{aligned}
$$

Therefore (ii) is true when $x = z'$, and so $z' \in M$.

Therefore, by **1c**, M contains every whole number, and so (i) and (ii) are true for every x and y.

We may now talk about *the* operation θ for which (i) and (ii) are true; (v) and (vi) are also true for it.

Definition (of *product* of whole numbers): The number $\theta(x, y)$ just defined is the *product* of x and y, and will be written as $x{\cdot}y$. (i), (ii), (v), and (vi) become

a. $x{\cdot}0 = 0 = 0{\cdot}x$ for every x

and

b. $x{\cdot}y' = x{\cdot}y + x$ and $x'{\cdot}y = x{\cdot}y + y$ for every x and y.

Note: $x{\cdot}y + x$ is short for $(x{\cdot}y) + x$, not for $x{\cdot}(y + x)$.

10. Theorem: $0'{\cdot}x = x$ *for every* x.

Proof:

$$\begin{aligned} 0'{\cdot}x &= 0{\cdot}x + x && \text{by } \mathbf{9b} \\ &= 0 + x && \text{by } \mathbf{9a} \\ &= x && \text{by } \mathbf{3a}. \end{aligned}$$

11. Theorem: *If* $x{\cdot}y = 0$, *then either* $x = 0$ *or* $y = 0$.

Proof: If $x{\cdot}y = 0$ and $y \neq 0$, then $y = u'$ for some u, by **2**. Then $0 = x{\cdot}y = x{\cdot}u' = x{\cdot}u + x$, by **9b**. Therefore $x = 0$, by **7**.

12. Theorem: $x{\cdot}(y + z) = x{\cdot}y + x{\cdot}z$ *for every* x, y, *and* z.

Proof: Let M be the set of z for which $x{\cdot}(y + z) = x{\cdot}y + x{\cdot}z$ for every x and y.

$$\begin{aligned} x{\cdot}(y + 0) &= x{\cdot}y && \text{by } \mathbf{3a} \\ &= x{\cdot}y + 0 && \text{by } \mathbf{3a} \\ &= x{\cdot}y + x{\cdot}0 && \text{by } \mathbf{9a}. \end{aligned}$$

Therefore $0 \in M$.

If $z \in M$, then

$$\begin{aligned} x{\cdot}(y + z') &= x{\cdot}(y + z)' && \text{by } \mathbf{3b} \\ &= x{\cdot}(y + z) + x && \text{by } \mathbf{9b} \\ &= (x{\cdot}y + x{\cdot}z) + x && \text{because } z \in M \\ &= x{\cdot}y + (x{\cdot}z + x) && \text{by } \mathbf{4} \\ &= x{\cdot}y + x{\cdot}z' && \text{by } \mathbf{9b}. \end{aligned}$$

Therefore $z' \in M$; and so, by **1c**, the theorem is true.

13. Theorem: $(x \cdot y) \cdot z = x \cdot (y \cdot z)$ *for every* x, y, *and* z.

Proof: Let M be the set of z for which $(x \cdot y) \cdot z = x \cdot (y \cdot z)$ for every x and y.

$$(x \cdot y) \cdot 0 = 0 = x \cdot 0 = x \cdot (y \cdot 0) \qquad \text{by } \mathbf{9a}.$$

Therefore $0 \in M$.

If $z \in M$, then

$$\begin{aligned} (x \cdot y) \cdot z' &= (x \cdot y) \cdot z + x \cdot y && \text{by } \mathbf{9b} \\ &= x \cdot (y \cdot z) + x \cdot y && \text{because } z \in M \\ &= x \cdot (y \cdot z + y) && \text{by } \mathbf{12} \\ &= x \cdot (y \cdot z') && \text{by } \mathbf{9b}. \end{aligned}$$

Therefore $z' \in M$; and so, by **1c**, the theorem is true.

14. Theorem: $x \cdot y = y \cdot x$ *for every* x *and* y.

Proof: Let M be the set of x for which $x \cdot y = y \cdot x$ for every y.

$$0 \cdot y = y \cdot 0 \quad \text{by } \mathbf{9a}.$$

Therefore $0 \in M$.

If $x \in M$, then

$$\begin{aligned} x' \cdot y &= x \cdot y + y && \text{by } \mathbf{9b} \\ &= y \cdot x + y && \text{because } x \in M \\ &= y \cdot x' && \text{by } \mathbf{9b}. \end{aligned}$$

Therefore $x' \in M$; and so, by **1c**, the theorem is true.

15. Theorem: $(x + y) \cdot z = x \cdot z + y \cdot z$ *for every* x, y, *and* z.

Proof:

$$\begin{aligned} (x + y) \cdot z &= z \cdot (x + y) && \text{by } \mathbf{14} \\ &= z \cdot x + z \cdot y && \text{by } \mathbf{12} \\ &= x \cdot z + y \cdot z && \text{by } \mathbf{14}. \end{aligned}$$

16. Theorem: *If* $x \cdot z = x \cdot y$ *and* $x \neq 0$, *then* $z = y$.

Proof: We have either $y = u + z$ or $z = v + y$, by **8**. Suppose first that $y = u + z$. Then if $x \cdot z = x \cdot y$ we have

$$\begin{aligned} 0 + x \cdot z &= x \cdot z && \text{by } \mathbf{3a} \\ &= x \cdot y \\ &= x \cdot (u + z) \\ &= x \cdot u + x \cdot z && \text{by } \mathbf{12}. \end{aligned}$$

Therefore $x \cdot u = 0$, by **6**. Therefore, if $x \neq 0$, we have $u = 0$, by **11**. Then $y = 0 + z = z$, by **3a**. The proof is similar if, instead of $y = u + z$, we have $z = v + y$.

17. Lemma (to **C21**): *$a \cdot c + b \cdot d = a \cdot d + b \cdot c$, if and only if $a = b$ or $c = d$.*

Proof:

I. Suppose that $a \cdot c + b \cdot d = a \cdot d + b \cdot c$. By **8**, either $a = u + b$ or $b = v + a$. If

(i) $$a = u + b,$$

then $$(u + b) \cdot c + b \cdot d = (u + b) \cdot d + b \cdot c$$

Therefore $(u \cdot c + b \cdot c) + b \cdot d = (u \cdot d + b \cdot d) + b \cdot c$ by **15**.

Therefore $(u \cdot c + b \cdot d) + b \cdot c = (u \cdot d + b \cdot d) + b \cdot c$ by **4** and **5**.

Therefore $u \cdot c + b \cdot d = u \cdot d + b \cdot d$ by **6**.

Therefore $u \cdot c = u \cdot d$ by **6**.

Therefore $u = 0$ or $c = d$ by **16**.

If $u = 0$, then $a = b$ by **3a** and (i).

Similarly, we prove that if $b = v + a$, then $a = b$ or $c = d$.

II. If $a = b$, then

$$\begin{aligned} a \cdot c + b \cdot d &= b \cdot c + a \cdot d \\ &= a \cdot d + b \cdot c \qquad \text{by } \mathbf{5}. \end{aligned}$$

And if $c = d$, then $a \cdot c + b \cdot d = a \cdot d + b \cdot c$.

18. Lemma (to **C 16**): *If $a + q = b + p$ and $c + s = d + r$, then $(a \cdot c + b \cdot d) + (p \cdot s + q \cdot r) = (a \cdot d + b \cdot c) + (p \cdot r + q \cdot s)$.*

Proof: By **8**, either $p = u + q$ or $q = v + p$.

If $$p = u + q$$

then
$$\begin{aligned} a + q &= b + p \\ &= b + (u + q) \\ &= (b + u) + q \qquad \text{by } \mathbf{4}. \end{aligned}$$

Therefore $a = b + u$ by **6**.

Now
$$\begin{aligned} u \cdot c + u \cdot s &= u \cdot (c + s) && \text{by } \mathbf{12} \\ &= u \cdot (d + r) && \text{because } c + s = d + r \end{aligned}$$

(i) $$= u \cdot d + u \cdot r \qquad \text{by } \mathbf{12}.$$

Then $(a{\cdot}c + b{\cdot}d) + (p{\cdot}s + q{\cdot}r)$

$$
\begin{aligned}
&= ((b + u){\cdot}c + b{\cdot}d) + ((u + q){\cdot}s + q{\cdot}r) && \\
&= ((b{\cdot}c + u{\cdot}c) + b{\cdot}d) + ((u{\cdot}s + q{\cdot}s) + q{\cdot}r) && \text{by } \mathbf{15} \\
&= (u{\cdot}c + u{\cdot}s) + ((b{\cdot}c + b{\cdot}d) + (q{\cdot}s + q{\cdot}r)) && \text{by } \mathbf{4} \text{ and } \mathbf{5} \\
&= (u{\cdot}d + u{\cdot}r) + ((b{\cdot}c + b{\cdot}d) + (q{\cdot}s + q{\cdot}r)) && \text{by (i)} \\
&= ((b{\cdot}d + u{\cdot}d) + b{\cdot}c) + ((u{\cdot}r + q{\cdot}r) + q{\cdot}s) && \text{by } \mathbf{4} \text{ and } \mathbf{5} \\
&= ((b + u){\cdot}d + b{\cdot}c) + ((u + q){\cdot}r + q{\cdot}s) && \text{by } \mathbf{15} \\
&= (a{\cdot}d + b{\cdot}c) + (p{\cdot}r + q{\cdot}s).
\end{aligned}
$$

The proof is similar if, instead of $p = u + q$, we have $q = v + p$.

EXERCISES A

1. Prove that $0'' + 0'' = 0''{\cdot}0'' = 0''''$.
2. Prove that, for every x, $x' \neq x$.
3. Prove that if $y = x + y$, then $x = 0$.
4. Prove that if $x = y + u$ and $y = x + v$, then $x = y$.
5. Prove that if $x + u = y$ and $y + v = x'$, then either $u = 0$ or $v = 0$.
6. Prove that if X is a set of whole numbers, there is a number x in X such that, given any y in X, $y = x + u$ for some u.
7. Prove that if b is not 0 and a is any whole number, then there are whole numbers q and r and a non-zero whole number s such that

(i) $a = b{\cdot}q + r$

and (ii) $b = r + s$.

8*. Which of the theorems **2–16** remain true if, instead of obeying axioms **1a**, **b**, and **c**, the succession obeys only (i) **1a** and **1c**, (ii) **1b** and **1c**, (iii) **1c**? Which theorems become necessarily false if **1a** is false but **1b** and **1c** are true; which if **1b** is false but **1a** and **1c** are true?

CHAPTER **B**

Hemigroups and Groups

1. Definition (of *hemigroup* and *dyad*): A *hemigroup* is a set S of elements and an operation on S with the following properties:

a. $x * y = y * x$ for every x and y of S;

b. $(x * y) * z = x * (y * z)$ for every x, y, and z of S;

c. If $x * y = z * y$, then $x = z$;

and

d. There is an element e of S such that $e * e = e$.

For any ordered pair (a, b) of elements of the hemigroup S, denote by $a \mid b$ the set of all ordered pairs (x, y) for which $x * b = y * a$. Then $a \mid b$ is a *dyad* of S and $*$.

Notation: In **B1–15**, small italic letters will be used for elements of S, and Greek letters for dyads.

Note: The statement **d** does not preclude there being elements other than e satisfying the equation $x * x = x$. However, it is not hard to prove that if **a**, **b**, **c**, and **d** are true, then e is the only such element.*

2. *Notation*: $x * (y * z)$ and $(x * y) * z$ will be written as $x * y * z$ (see **1b**).

Theorem:

a. $x * y * z = x * z * y = y * x * z = y * z * x = z * y * x = z * x * y$

and

b. $(x * y) * (z * w) = (x * z) * (y * w)$.

Proof: By **1a** and **1b**.

* Hint: If $f * f = f$, consider $f * (f * e)$ and $f * (e * e)$.

3. Theorem: *The ordered pair (p, q) is in the dyad $p \mid q$ and in no other dyad.*

Proof: $(p, q) \in p \mid q$ by definition **1**, because $p * q = q * p$, by **1a**. If $(p, q) \in a \mid b$, then

(i) $$p * b = q * a$$

by definition **1**.

Now if $(x, y) \in a \mid b$, then

(ii) $$x * b = y * a$$

by definition **1**.

Therefore
$$\begin{aligned} x * q * a &= x * p * b && \text{by (i)} \\ &= x * b * p && \text{by } \mathbf{2} \\ &= y * a * p && \text{by (ii)} \\ &= y * p * a && \text{by } \mathbf{2}. \end{aligned}$$

Therefore $x * q = y * p$, by **1c**; and so, by definition **1**, $(x, y) \in p \mid q$. Thus we have proved that every element of $a \mid b$ is an element of $p \mid q$.

Now if $(x, y) \in p \mid q$, then

(iii) $$x * q = y * p \qquad \text{by definition } \mathbf{1}.$$

Then
$$\begin{aligned} x * b * p &= x * p * b && \text{by } \mathbf{2} \\ &= x * q * a && \text{by (i)} \\ &= y * p * a && \text{by (iii)} \\ &= y * a * p && \text{by } \mathbf{2}. \end{aligned}$$

Therefore, by **1c**, $x * b = y * a$.

Therefore, by definition **1**, $(x, y) \in a \mid b$.

Therefore every element of $p \mid q$ is an element of $a \mid b$; and so $p \mid q = a \mid b$.

4. Theorem: *$a \mid b = c \mid d$ if and only if $a * d = b * c$.*

Proof: By **3**, $a \mid b = c \mid d$ if and only if $(a, b) \in c \mid d$. By definition **1**, this is so if and only if $a * d = b * c$.

5. Lemma (to definition **6**): *If (a_1, b_1) and (a_2, b_2) are in the same dyad, and (g_1, h_1) and (g_2, h_2) are in the same dyad; then $(a_1 * g_1, b_1 * h_1)$ and $(a_2 * g_2, b_2 * h_2)$ are in the same dyad.*

Proof: $(a_2, b_2) \in a_2 \mid b_2$, by **3**; and so, if (a_1, b_1) and (a_2, b_2) are in the same dyad, then

(i) $$a_1 * b_2 = b_1 * a_2$$

by definition **1**. Similarly

(ii) $$g_1 * h_2 = h_1 * g_2.$$

Then $(a_1 * g_1) * (b_2 * h_2) = (a_1 * b_2) * (g_1 * h_2)$ by **2**

$= (b_1 * a_2) * (h_1 * g_2)$ by (i) and (ii)

$= (b_1 * h_1) * (a_2 * g_2)$ by **2**.

Therefore $(a_1 * g_1, b_1 * h_1) \in (a_2 * g_2) \mid (b_2 * h_2)$ by definition **1**.

But $(a_2 * g_2, b_2 * h_2) \in (a_2 * g_2) \mid (b_2 * h_2)$ by **3**.

Therefore these are in the same dyad.

6. Definition (of $*$ applied to dyads): $\xi * \eta$ is the dyad containing all $(a * g, b * h)$ where $(a, b) \in \xi$ and $(g, h) \in \eta$.

Note: That there is a dyad which contains them all follows from **5**. What we have done is to define an operation on dyads based on the operation $*$, and we have used the same sign, $*$, to denote the new operation. We shall always use the same sign for the operation on dyads as for the hemigroup operation from which it is deduced. (In Chapters **C** and **E** we shall meet hemigroups whose operations are addition and multiplication.)

7. Theorem: $p \mid q * r \mid s = (p * r) \mid (q * s)$.

Proof: $(p * r, q * s) \in (p * r) \mid (q * s)$, by **3**. But $(p, q) \in p \mid q$, and $(r, s) \in r \mid s$, by **3**. Therefore $(p * r) \mid (q * s)$ contains an element of the form $(a * g, b * h)$ where $(a, b) \in p \mid q$ and $(g, h) \in r \mid s$. Therefore, by **5**, it contains them all. Therefore, by definition **6**, it is $p \mid q * r \mid s$.

8. Theorem: $p \mid q = (r * p) \mid (r * q)$.

Proof: $p * r * q = q * r * p$ by **2**.

Therefore $(p, q) \in (r * p) \mid (r * q)$ by definition **1**.

Therefore $p \mid q = (r * p) \mid (r * q)$ by **3**.

9. Theorem: $\xi * \eta = \eta * \xi$.

Proof: Let ξ be $a \mid b$ and η be $c \mid d$. Then

$$\begin{aligned} \xi * \eta &= a \mid b * c \mid d \\ &= (a * c) \mid (b * d) && \text{by } \mathbf{7} \\ &= (c * a) \mid (d * b) && \text{by } \mathbf{1a} \\ &= c \mid d * a \mid b && \text{by } \mathbf{7} \\ &= \eta * \xi. \end{aligned}$$

10. Theorem $(\xi * \eta) * \zeta = \xi * (\eta * \zeta)$.

Proof: Let ξ be $a \mid b$, η be $c \mid d$, and ζ be $g \mid f$. Then

$$
\begin{aligned}
(\xi * \eta) * \zeta &= (a \mid b * c \mid d) * g \mid f \\
&= ((a * c) \mid (b * d)) * g \mid f && \text{by } \mathbf{7} \\
&= (a * c * g) \mid (b * d * f) && \text{by } \mathbf{7} \text{ and notation } \mathbf{2} \\
&= a \mid b * ((c * g) \mid (d * f)) && \text{by } \mathbf{7} \\
&= a \mid b * (c \mid d * g \mid f) && \text{by } \mathbf{7} \\
&= \xi * (\eta * \zeta).
\end{aligned}
$$

11. Theorem: *The correspondence $x \mid e \longleftrightarrow x$ between dyads of the form $x \mid e$ and S is a $*$-isomorphism.*

Proof: $x \mid e = y \mid e$ if and only if $x * e = e * y$, by **4**. This is so if and only if $x = y$, by **1a** and **1c**. Therefore the correspondence is a matching.

$$
\begin{aligned}
x \mid e * y \mid e &= (x * y) \mid (e * e) && \text{by } \mathbf{7} \\
&= (x * y) \mid e && \text{by } \mathbf{1d}.
\end{aligned}
$$

Therefore the correspondence is a $*$-isomorphism.

12. Definition (of ε): ε is $e \mid e$.

13. Theorem: $x \mid y = \varepsilon$ *if and only if* $x = y$.

Proof: $x \mid y = e \mid e$ if and only if $x * e = y * e$, by **4**. This is so if and only if $x = y$, by **1c**.

14. Theorem: $\varepsilon * \xi = \xi$, *for every* ξ.

Proof: Let ξ be $b \mid c$. Then
$$
\begin{aligned}
\varepsilon * \xi &= e \mid e * b \mid c \\
&= (e * b) \mid (e * c) && \text{by } \mathbf{7} \\
&= b \mid c && \text{by } \mathbf{8} \\
&= \xi.
\end{aligned}
$$

15. Theorem: *For each dyad α there is a dyad $\bar{\alpha}$ such that $\alpha * \bar{\alpha} = \varepsilon$. Moreover, if α is $a \mid b$, then $b \mid a$ is such a dyad.*

Proof:
$$
\begin{aligned}
a \mid b * b \mid a &= (a * b) \mid (b * a) && \text{by } \mathbf{7} \\
&= (a * b) \mid (a * b) && \text{by } \mathbf{1a} \\
&= \varepsilon && \text{by } \mathbf{13}.
\end{aligned}
$$

16. Definition (of *commutative group*): A *commutative group* is a set Γ and an operation $*$ on it with the following properties.

a. $(\xi * \eta) * \zeta = \xi * (\eta * \zeta)$ for every ξ, η, and ζ of Γ.

b. $\xi * \eta = \eta * \xi$ for every ξ and η of Γ.

c. There is a ε of Γ such that $\varepsilon * \xi = \xi$ for every ξ of Γ.

d. For each η of Γ there is a $\bar{\eta}$ of Γ such that $\eta * \bar{\eta} = \varepsilon$.

Note: From **b** and **d** we have

e. For each η of Γ there is a $\bar{\eta}$ of Γ such that $\bar{\eta} * \eta = \varepsilon$.

Notation: In **16–21** small Greek letters will denote elements of a commutative group. Γ will denote the set of elements of the group and $*$ the operation of the group.

17. Theorem: *If α and β are elements of Γ, there is a ξ for which $\alpha * \xi = \beta$; and if $\alpha * \eta = \alpha * \xi$, then $\eta = \xi$.*

Proof: By **16d** there is a $\bar{\alpha}$ such that $\alpha * \bar{\alpha} = \varepsilon$. Let ξ be $\bar{\alpha} * \beta$. Then

$$\begin{aligned} \alpha * \xi &= (\alpha * \bar{\alpha}) * \beta && \text{by } \mathbf{16a} \\ &= \varepsilon * \beta \\ &= \beta && \text{by } \mathbf{16c}. \end{aligned}$$

On the other hand, if $\alpha * \eta = \alpha * \xi$, then

$$\begin{aligned} \eta &= \varepsilon * \eta && \text{by } \mathbf{16c} \\ &= (\bar{\alpha} * \alpha) * \eta && \text{by } \mathbf{16e} \\ &= \bar{\alpha} * (\alpha * \eta) && \text{by } \mathbf{16a} \\ &= \bar{\alpha} * (\alpha * \xi) \\ &= (\bar{\alpha} * \alpha) * \xi && \text{by } \mathbf{16a} \\ &= \varepsilon * \xi && \text{by } \mathbf{16e} \\ &= \xi && \text{by } \mathbf{16c}. \end{aligned}$$

18. Theorem: *If $\iota * \xi = \xi$, then $\iota = \varepsilon$; if $\alpha * \kappa = \varepsilon$, then $\kappa = \bar{\alpha}$.*

Proof:

$$\begin{aligned} \xi * \iota &= \iota * \xi && \text{by } \mathbf{16b} \\ &= \xi \\ &= \varepsilon * \xi && \text{by } \mathbf{16c} \\ &= \xi * \varepsilon && \text{by } \mathbf{16b}. \end{aligned}$$

Therefore $\iota = \varepsilon$ by **17.**

And $\alpha * \kappa = \varepsilon = \alpha * \bar{\alpha}$ by **16d.**

Therefore $\kappa = \bar{\alpha}$ by **17.**

Note: We may now talk about *the* element ε for which **16c** is true, and *the* element $\bar{\eta}$ for which **16d** is true.

19. Theorem: $\bar{\bar{\eta}} = \eta.$

Proof: $\bar{\eta} * \eta = \varepsilon$ by **16e**.

Therefore $\eta = \bar{\bar{\eta}}$ by **18**.

20. Theorem: $\bar{\varepsilon} = \varepsilon.$

Proof: $\varepsilon * \varepsilon = \varepsilon$ by **16c**.

Therefore $\varepsilon = \bar{\varepsilon}$ by **18**.

21. Theorem: $\overline{\xi * \eta} = \bar{\xi} * \bar{\eta}.$

Proof:

$$
\begin{aligned}
(\xi * \eta) * (\bar{\xi} * \bar{\eta}) &= (\xi * \bar{\xi}) * (\eta * \bar{\eta}) && \text{by 16a and b} \\
&= \varepsilon * \varepsilon && \text{by 16d} \\
&= \varepsilon && \text{by 16c.}
\end{aligned}
$$

Therefore $\bar{\xi} * \bar{\eta} = \overline{\xi * \eta}$ by **18**.

22. Theorem: *The set of dyads defined in* **1** *and the operation defined in* **6** *form a commutative group.*

Proof: **16a, b, c,** and **d** follow from **10, 9, 14,** and **15** respectively:

$$\overline{x \mid y} \text{ is } y \mid x.$$

EXERCISES B

1*. Which of the following are hemigroups and which are commutative groups under the given operations?

(i) The whole numbers greater than 1 (addition).
(ii) The whole numbers 1, 2, . . . m ($x * y$ is the remainder on dividing the ordinary sum $x + y$ by m).
(iii) The whole numbers 1, 2, . . . $m - 1$ ($x * y$ is the remainder on dividing the product $x{\cdot}y$ by m).
(iv) The integers ($x * y = x + y - x{\cdot}y$).
(v) The integers other than 1 ($x * y = x + y - x{\cdot}y$).
(vi) The whole numbers not less than k ($x * y = x + y - k$).

2. $f(x)$ is $1/x$, and $g(x)$ is $1 - x$. The operation $*$ is defined as follows: if θ and ϕ are any functions, then $\theta * \phi$ is the function ψ for which

$$\psi(x) = \theta(\phi(x)) \text{ for every } x$$

[For example, $(f * g)(x) = f(g(x)) = f(1 - x) = 1/(1 - x)$.]

Prove that with this operation the elements f, g, $f * g$, $f * f$, $g * f$, and $g * f * g$ form a non-commutative group: that is, **16a**, **c**, and **e** are satisfied, but not **16b**.

3. Prove that if a hemigroup S happens also to be a group (under the same operation), and if we form a group of dyads from it, then this group is isomorphic to S.

4. Prove that a hemigroup with a finite number of elements is a group.

5. (i) Prove that $(y * x) \mid y = (z * x) \mid z$.
   (ii) Prove that, for each a of S, $(a * x) \mid a \longleftrightarrow x$ is an isomorphism.
   (iii) Prove that if we have a system T for which **1a**, **b**, and **c** are true but not **1d**, then the theorems in **B** still remain true if we restate **11** in the form "there is an isomorphism between certain dyads and the elements of T".

6. Show how to prove **17**, **18**, **19**, and **20** without using **16b** and using **16e** in place of **16d**. Prove in this way a theorem as nearly like **21** as you can.

CHAPTER C

Integers

1. Theorem: *The set of whole numbers, with the operation of addition, is a hemigroup.*

Proof: **B1a, b, c,** and **d** follow from **A5, 4, 6,** and **3a**, with 0 playing the part of e.

Definition (of *integer*): The dyads of the whole numbers and addition are *integers*.

Notation: In Chapter **C** (as also in **E** and **G**), Greek letters will denote integers. Roman letters (as in **A**) will denote whole numbers. We shall, when our hemigroup is the hemigroup of whole numbers, write $x - y$ instead of $x \mid y$, $-\alpha$ instead of $\bar{\alpha}$, $\xi + \eta$ instead of $\xi * \eta$, and o instead of ε; that is, $o = 0 - 0$.

Applying the results of Chapter **B** to the integers, we have immediately:

2. *$p - q = r - s$ if and only if $p + s = q + r$, and $p - q = o$ if and only if $p = q$* by **B4** and **13.**

3. $(p - q) + (r - s) = (p + r) - (q + s)$ by **B7.**

4. $p - q = (r + p) - (r + q)$ by **B8.**

5. *The integers, with the operation of addition, form a commutative group* by **B22.**

6. $\xi + \eta = \eta + \xi$ by **B9.**

7. $(\xi + \eta) + \zeta = \xi + (\eta + \zeta)$ by **B10.**

8. *For any α and β, there is a ξ such that $\alpha + \xi = \beta$; and if $\alpha + \eta = \alpha + \xi$, then $\eta = \xi$* by **B17.**

9. *The correspondence $x - 0 \longleftrightarrow x$ between integers of the form $x - 0$ and the whole numbers is an addition-isomorphism* by **B11.**

10. $\xi + o = o + \xi = \xi$ by **B9** and **14**.

11. $\xi + (-\xi) = o$ by **B15**.

12. $-(x - y) = y - x$ by **B15**.

13. $-(-\xi) = \xi$ by **B19**.

14. $-o = o$ by **B20**.

15. $-(\xi + \eta) = -\xi + (-\eta)$ by **B21**.

16. Lemma (to definition **C17**): *If*

(i) $$a - b = p - q$$

and

(ii) $$c - d = r - s$$

then

(iii) $$(a{\cdot}c + b{\cdot}d) - (a{\cdot}d + b{\cdot}c) = (p{\cdot}r + q{\cdot}s) - (p{\cdot}s + q{\cdot}r).$$

Proof: If (i) is true, then $a + q = b + p$ by **C2**.
If (ii) is true, then $c + s = d + r$ by **C2**.

Then $(a{\cdot}c + b{\cdot}d) + (p{\cdot}s + q{\cdot}r) = (a{\cdot}d + b{\cdot}c) + (p{\cdot}r + q{\cdot}s)$ by **A18**.

Then (iii) is true by **C2**.

17. Definition (of *product* of integers): $\xi{\cdot}\eta$ is

$(a{\cdot}c + b{\cdot}d) - (a{\cdot}d + b{\cdot}c)$, where $a - b = \xi$ and $c - d = \eta$.

Note: By **C16**, this integer is the same no matter which a, b, c, and d are chosen, as long as $a - b = \xi$ and $c - d = \eta$.

18. Theorem: $\xi{\cdot}\eta = \eta{\cdot}\xi$.

Proof: Let ξ be $a - b$ and η be $c - d$.

Then
$$\begin{aligned} \xi{\cdot}\eta &= (a{\cdot}c + b{\cdot}d) - (a{\cdot}d + b{\cdot}c) && \text{by definition } \mathbf{C17} \\ &= (c{\cdot}a + d{\cdot}b) - (c{\cdot}b + d{\cdot}a) && \text{by } \mathbf{A14} \text{ and } \mathbf{5} \\ &= \eta{\cdot}\xi && \text{by definition } \mathbf{C17}. \end{aligned}$$

19. Theorem: $(\xi{\cdot}\eta){\cdot}\zeta = \xi{\cdot}(\eta{\cdot}\zeta)$.

Proof: Let ξ be $a - b$, η be $c - d$, and ζ be $g - f$.

Then $(\xi\cdot\eta)\cdot\zeta$

$$= ((a\cdot c + b\cdot d) - (a\cdot d + b\cdot c))\cdot(g - f) \qquad \text{by definition } \mathbf{C17}$$

$$= ((a\cdot c + b\cdot d)\cdot g + (a\cdot d + b\cdot c)\cdot f) - ((a\cdot c + b\cdot d)\cdot f + (a\cdot d + b\cdot c)\cdot g) \qquad \text{by definition } \mathbf{C17}$$

$$= (((a\cdot c)\cdot g + (b\cdot d)\cdot g) + ((a\cdot d)\cdot f + (b\cdot c)\cdot f)) - (((a\cdot c)\cdot f + (b\cdot d)\cdot f) + ((a\cdot d)\cdot g + (b\cdot c)\cdot g)) \qquad \text{by } \mathbf{A15}$$

$$= ((a\cdot(c\cdot g) + a\cdot(d\cdot f)) + (b\cdot(c\cdot f) + b\cdot(d\cdot g))) - ((a\cdot(c\cdot f) + a\cdot(d\cdot g)) + (b\cdot(c\cdot g) + b\cdot(d\cdot f))) \qquad \text{by } \mathbf{A4}, \mathbf{5} \text{ and } \mathbf{13}$$

$$= (a\cdot(c\cdot g + d\cdot f) + b\cdot(c\cdot f + d\cdot g)) - (a\cdot(c\cdot f + d\cdot g) + b\cdot(c\cdot g + d\cdot f)) \qquad \text{by } \mathbf{A12}$$

$$= (a - b)\cdot((c\cdot g + d\cdot f) - (c\cdot f + d\cdot g)) \qquad \text{by definition } \mathbf{C17}$$

$$= \xi\cdot((c - d)\cdot(g - f)) \qquad \text{by definition } \mathbf{C17}$$

$$= \xi\cdot(\eta\cdot\zeta).$$

20. Theorem: $\xi\cdot(\eta + \zeta) = \xi\cdot\eta + \xi\cdot\zeta$.

Proof: Let ξ be $a - b$, η be $c - d$, and ζ be $g - f$.

Then $\xi\cdot(\eta + \zeta) = (a - b)\cdot((c - d) + (g - f))$

$$= (a - b)\cdot((c + g) - (d + f)) \qquad \text{by } \mathbf{C3}$$

$$= (a\cdot(c + g) + b\cdot(d + f)) - (a\cdot(d + f) + b\cdot(c + g)) \qquad \text{by definition } \mathbf{C17}$$

$$= ((a\cdot c + a\cdot g) + (b\cdot d + b\cdot f)) - ((a\cdot d + a\cdot f) + (b\cdot c + b\cdot g)) \qquad \text{by } \mathbf{A12}$$

$$= ((a\cdot c + b\cdot d) + (a\cdot g + b\cdot f)) - ((a\cdot d + b\cdot c) + (a\cdot f + b\cdot g)) \qquad \text{by } \mathbf{A4} \text{ and } \mathbf{5}$$

$$= ((a\cdot c + b\cdot d) - (a\cdot d + b\cdot c)) + ((a\cdot g + b\cdot f) - (a\cdot f + b\cdot g)) \qquad \text{by } \mathbf{C3}$$

$$= (a - b)\cdot(c - d) + (a - b)\cdot(g - f) \qquad \text{by definition } \mathbf{C17}$$

$$= \xi\cdot\eta + \xi\cdot\zeta.$$

21. Theorem: $\xi\cdot\eta = o$ *if and only if* $\xi = o$ *or* $\eta = o$.

Proof: Let ξ be $a - b$ and η be $c - d$. Then

$$\xi\cdot\eta = (a - b)\cdot(c - d)$$

$$= (a\cdot c + b\cdot d) - (a\cdot d + b\cdot c) \qquad \text{by definition } \mathbf{C17}.$$

This is equal to o if and only if $a\cdot c + b\cdot d = a\cdot d + b\cdot c$, by **C2**. But this is true if and only if $a = b$ or $c = d$, by **A17**; that is, if and only if $\xi = o$ or $\eta = o$, by **C2**.

22. Theorem: *If* $\xi \cdot \eta = \zeta \cdot \eta$ *and* $\eta \neq o$, *then* $\xi = \zeta$.

Proof:
$$\begin{aligned}
(\xi + (-\zeta))\cdot\eta &= \xi\cdot\eta + (-\zeta)\cdot\eta && \text{by } \mathbf{C18} \text{ and } \mathbf{20}\\
&= \zeta\cdot\eta + (-\zeta)\cdot\eta\\
&= (\zeta + (-\zeta))\cdot\eta && \text{by } \mathbf{C18} \text{ and } \mathbf{20}\\
&= o\cdot\eta && \text{by } \mathbf{C11}\\
&= o && \text{by } \mathbf{C21}.
\end{aligned}$$

But $\eta \neq o$.

Therefore
$$\begin{aligned}
\xi + (-\zeta) &= o && \text{by } \mathbf{C21}.\\
&= \zeta + (-\zeta) && \text{by } \mathbf{C11}.
\end{aligned}$$

Therefore $\xi = \zeta$ by **C6** and **C8**.

23. Theorem: *The correspondence* $x - 0 \longleftrightarrow x$ *between integers of the form* $x - 0$ *and the whole numbers is a multiplication-isomorphism.*

Proof:
$$\begin{aligned}
(x - 0)\cdot(y - 0) &= (x\cdot y + 0\cdot 0) - (x\cdot 0 + 0\cdot y) && \text{by definition } \mathbf{C17}\\
&= x\cdot y - 0 && \text{by } \mathbf{A9a} \text{ and } \mathbf{3a}.
\end{aligned}$$

Note: We have seen in **C9** that the correspondence is an addition-isomorphism.

24. Definition (of *positive* integer): An integer of the form $x - 0$ is *positive*.

Note: the positive integers are thus the integers which correspond to the whole numbers in isomorphism **C23**.

25. Definition (of o'): o' is $0' - 0$.

Note: o and o' are positive, by definition **C24**.

26. Theorem: $o'\cdot\xi = \xi$.

Proof: Let ξ be $x - y$. Then
$$\begin{aligned}
o'\cdot\xi &= (0' - 0)\cdot(x - y) && \text{by definition } \mathbf{C25}\\
&= (0'\cdot x + 0\cdot y) - (0'\cdot y + 0\cdot x) && \text{by definition } \mathbf{C17}\\
&= x - y && \text{by } \mathbf{A10}, \mathbf{9a}, \text{ and } \mathbf{3a}\\
&= \xi.
\end{aligned}$$

27. Theorem: *Either* ξ *or* $-\xi$ *is positive.*

Proof: Let ξ be $a - b$. By **A8**, either $a = u + b$ or $b = u + a$ for some u.

If $a = u + b$, then $\xi = a - b = u - 0$, by **C2**, **A5**, and **A3a**, and so is positive, by definition **C24**.

If $b = u + a$, then $\xi = a - b = 0 - u$, by **C2**, **A5**, and **A3a**. Therefore $-\xi = u - 0$, by **C12**, which is positive, by definition **C24**.

28. Theorem: $(-\xi)\cdot\eta = -(\xi\cdot\eta) = \xi\cdot(-\eta)$.

Proof: Let ξ be $a - b$ and η be $c - d$. Then

$$
\begin{aligned}
(-\xi)\cdot\eta &= (b - a)\cdot(c - d) && \text{by } \mathbf{C12} \\
&= (b\cdot c + a\cdot d) - (b\cdot d + a\cdot c) && \text{by definition } \mathbf{C17} \\
&= -((b\cdot d + a\cdot c) - (b\cdot c + a\cdot d)) && \text{by } \mathbf{C12} \\
&= -((a\cdot c + b\cdot d) - (a\cdot d + b\cdot c)) && \text{by } \mathbf{A5} \\
&= -((a - b)\cdot(c - d)) && \text{by definition } \mathbf{C17} \\
\text{(i)} \qquad &= -(\xi\cdot\eta).
\end{aligned}
$$

Then

$$
\begin{aligned}
\xi\cdot(-\eta) &= (-\eta)\cdot\xi && \text{by } \mathbf{C18} \\
&= -(\eta\cdot\xi) && \text{by (i)} \\
&= -(\xi\cdot\eta) && \text{by } \mathbf{C18}.
\end{aligned}
$$

29. Theorem: $(-\xi)\cdot(-\eta) = \xi\cdot\eta$.

Proof:

$$
\begin{aligned}
(-\xi)\cdot(-\eta) &= -(\xi\cdot(-\eta)) && \text{by } \mathbf{C28} \\
&= -(-(\xi\cdot\eta)) && \text{by } \mathbf{C28} \\
&= \xi\cdot\eta && \text{by } \mathbf{C13}.
\end{aligned}
$$

30. Theorem: *If ξ and η are positive, then so are $\xi + \eta$ and $\xi\cdot\eta$.*

Proof: By **C24**, ξ and η correspond to whole numbers, a and b, say, in isomorphism **C23**. Then $\xi\cdot\eta$ and $\xi + \eta$ correspond to $a\cdot b$ and $a + b$, which are whole numbers. Therefore $\xi\cdot\eta$ and $\xi + \eta$ are positive.

31. Theorem: *If ξ and $-\xi$ are both positive, then $\xi = o$.*

Proof: If ξ and $-\xi$ are positive, then $\xi = x - 0$ and $-\xi = y - 0$, by definition **C24**.

$$
\begin{aligned}
\text{Then} \qquad -\xi &= 0 - x && \text{by } \mathbf{C12} \\
\text{that is} \qquad y - 0 &= 0 - x. \\
\text{Therefore} \qquad y + x &= 0 + 0 && \text{by } \mathbf{C2} \\
&= 0 && \text{by } \mathbf{A3a}. \\
\text{Therefore} \qquad x &= 0 && \text{by } \mathbf{A7}. \\
\text{Therefore} \qquad \xi &= 0 - 0 = o && \text{by notation } \mathbf{C1}.
\end{aligned}
$$

32. Lemma (to **G6**): *If $\xi\cdot\beta = \gamma$, $\beta \neq o$, and β and γ are positive, then ξ is positive.*

Proof: If ξ is not positive, then $-\xi$ is positive, by **C27**.

Then
$$-\gamma = -(\xi\cdot\beta)$$
$$= (-\xi)\cdot\beta \qquad \text{by } \mathbf{C28}$$
$$\text{which is positive} \qquad \text{by } \mathbf{C30}.$$

Therefore $\gamma = o$, by **C31**; i.e. $\xi\cdot\beta = o$.
Therefore $\xi = o$, by **C21** and because $\beta \neq o$.
But o is positive, by note **C25**. Therefore ξ cannot *not* be positive.

33. Lemma (to **G8**): *If β is positive and not equal to o, and if α is positive, then $\alpha\cdot\beta + (-\alpha)$ is positive.*

Proof: Let $\beta = b - 0$ and $\alpha = a - 0$. Then $b \neq 0$. Therefore $b = u'$ for some u, by **A2**. Let $\phi = u - 0$; ϕ will then be positive, by **C24**.

Then $\qquad a\cdot b = a\cdot u' = a\cdot u + a \qquad$ by **A9b**.
Therefore $\qquad \alpha\cdot\beta = \alpha\cdot\phi + \alpha \qquad$ by **C23** and **C9**.
Therefore
$$\alpha\cdot\beta + (-\alpha) = (\alpha\cdot\phi + \alpha) + (-\alpha)$$
$$= \alpha\cdot\phi + (\alpha + (-\alpha)) \qquad \text{by } \mathbf{C7}$$
$$= \alpha\cdot\phi + o \qquad \text{by } \mathbf{C11}$$
$$= \alpha\cdot\phi \qquad \text{by } \mathbf{C10}$$
$$\text{which is positive} \qquad \text{by } \mathbf{C30}.$$

EXERCISES C

1. Prove that $o' + o' = o''$, where $o'' = 0'' - 0$.
2. Prove that $o'' \neq o'$.
3. Prove that if we define $\xi^\times$ to be $\xi + (-o')$, then the set of integers ξ for which $-\xi$ is positive form a succession if we take $\xi^\times$ as the successor of ξ. Prove that addition of elements of this succession, defined as in **A3**, is the same as addition of the elements considered as integers, but that the two definitions of multiplication are different.
4. Prove that the correspondence $0 - x \longleftrightarrow x$ between integers of the form $0 - x$ and whole numbers is an addition-isomorphism but not a multiplication-isomorphism.
5. Prove that if α is an integer, there is an integer β such that $\alpha + \beta$ is positive.
6. Prove that if $\beta \neq o$ and α is any integer, there is a γ such that $\beta\cdot\gamma + (-\alpha)$ is positive.

CHAPTER D

Fields

1. Definition (of *field*): A *field* consists of a set F and two operations on it (which we shall denote by $+$ and $\cdot$) with the following properties:

a. $(x + y) + z = x + (y + z)$ for every x, y, and z of F.

b. $x + y = y + x$ for every x and y of F.

c. There is a o of F such that $o + x = x$ for every x of F.

d. For each a of F there is a $-a$ of F such that $a + (-a) = o$.

e. $(x\cdot y)\cdot z = x\cdot(y\cdot z)$ for every x, y, and z of F.

f. $x\cdot y = y\cdot x$ for every x and y of F.

g. There is a i of F such that $i\cdot x = x$ for every x of F; and $i \neq o$.

h. For each a, other than o, of F there is a a^- of F such that $a\cdot a^- = i$.

k. $x\cdot(y + z) = x\cdot y + x\cdot z$, for every x, y, and z of F.

Notation: In **D**, small italic letters will denote elements of a field F.

2. Theorem: *The set F and the operation $+$ form a commutative group.*

Proof: **B16a, b, c,** and **d** follow from **D1a, b, c,** and **d.**

Then from **B17, 18, 19, 20,** and **21** we have:

a. For any a and b there is a x such that $a + x = b$; and if $a + x = a + y$, then $x = y$.

b. If $x + a = a$, then $x = o$.

c. If $a + x = o$, then $x = -a$.

d. $-(-x) = x$.

e. $-o = o$.

f. $-(x + y) = -x + (-y)$.

3. Theorem: *$x \cdot o = o \cdot x = o$ for every x of F.*

Proof:

$$x \cdot o + o = x \cdot o \quad \text{by } \mathbf{D1b} \text{ and } \mathbf{c}$$
$$= x \cdot (o + o) \quad \text{by } \mathbf{D1c}$$
$$= x \cdot o + x \cdot o \quad \text{by } \mathbf{D1k}.$$

Therefore

$$o = x \cdot o \quad \text{by } \mathbf{D2a}.$$

Then

$$x \cdot o = o \cdot x \quad \text{by } \mathbf{D1f}.$$

4. Theorem: *If $x \cdot y = o$, then $x = o$ or $y = o$.*

Proof: If $y \neq o$, we have

$$x = x \cdot i \quad \text{by } \mathbf{D1f} \text{ and } \mathbf{g}$$
$$= x \cdot (y \cdot y^-) \quad \text{by } \mathbf{D1h}$$
$$= (x \cdot y) \cdot y^- \quad \text{by } \mathbf{D1e}$$
$$= o \cdot y^-$$
$$= o \quad \text{by } \mathbf{D3}.$$

5. Theorem: *The operation $\cdot$ and the set K of elements other than o of F form a commutative group.*

Proof: If x and y are in K then neither is equal to o, and so $x \cdot y \neq o$, by **D4**. Therefore $x \cdot y$ is in K. i is in K, by **D1g**.

Then **D1e, f, g,** and **h** imply that K and $\cdot$ have properties **B16a, b, c,** and **d.**

Then from **B18, 19, 20,** and **21** we have:

a. If $x \cdot a = a$ and $a \neq o$, then $x = i$.

b. If $a \cdot x = i$, then $x = a^-$. (*Note*: we automatically have $a \neq o$, by **D3** and **D1g.**)

c. $(x^-)^- = x$.

d. $i^- = i$.

e. $(x \cdot y)^- = x^- \cdot y^-$.

6. Theorem: *If $a \neq o$, then*

(i)
$$a \cdot x = b$$

if and only if $x = a^- \cdot b$.

Proof: If $b \neq o$, this follows from **D5** and **B17**.
If $b = o$, then (i) is true if and only if $x = o$, by **D3** and **4**, and, in this case, $a^- \cdot b = o$, by **D3**.

7. Theorem: *$(x + y)\cdot z = x\cdot z + y\cdot z$ for every x, y, and z of F.*

Proof:

$$
\begin{aligned}
(x + y)\cdot z &= z\cdot(x + y) && \text{by } \mathbf{D1f}\\
&= z\cdot x + z\cdot y && \text{by } \mathbf{D1k}\\
&= x\cdot z + y\cdot z && \text{by } \mathbf{D1f}.
\end{aligned}
$$

8. Theorem: *$x\cdot(-y) = (-x)\cdot y = -(x\cdot y)$ for every x and y of F.*

Proof:

$$
\begin{aligned}
x\cdot y + x\cdot(-y) &= x\cdot(y + (-y)) && \text{by } \mathbf{D1k}\\
&= x\cdot o && \text{by } \mathbf{D1d}\\
&= o && \text{by } \mathbf{D3}.
\end{aligned}
$$

Therefore $x\cdot(-y) = -(x\cdot y)$ by **D2c.**
Then, using **D1f**, $(-x)\cdot y = y\cdot(-x) = -(y\cdot x) = -(x\cdot y)$.

9. Theorem: *$(-x)\cdot(-y) = x\cdot y$ for every x and y of F.*

Proof:

$$
\begin{aligned}
(-x)\cdot(-y) &= -((-x)\cdot y) && \text{by } \mathbf{D8}\\
&= -(-(x\cdot y)) && \text{by } \mathbf{D8}\\
&= x\cdot y && \text{by } \mathbf{D2d}.
\end{aligned}
$$

EXERCISES D

1. Prove that the set consisting of the elements a and b with the rules $a + a = b + b = a$, $a + b = b + a = b$, $a\cdot a = a\cdot b = b\cdot a = a$, and $b\cdot b = b$ is a field.

2. Construct a field with just three elements.

3*. Show that there are fields with the following property:

$$i + i + i + \dots i = o.$$

Prove that, in each such field, the smallest number of i's which sum to zero must be prime. Prove that a field which does not have this property contains a set of elements isomorphic to the integers.

4. Prove **D1b** from the other axioms in **D1.**

CHAPTER E

Rational Numbers

1. Theorem: *The set of non-zero integers, with the operation of multiplication, is a hemigroup.*

Proof: **B1a**, **b**, **c**, and **d** follow from **C18**, **19**, **22**, and **26**, with o' playing the part of e.

Notation: We shall, when our hemigroup is the hemigroup of non-zero integers, write dyads in the form α/β, and shall write $\boldsymbol{1}$ instead of ε for the element o'/o'. If $\boldsymbol{x} = \alpha/\beta$, we shall write $\boldsymbol{x}^-$ for β/α (see **B15**).

Definitions:

a. (of $\boldsymbol{0}$): $\boldsymbol{0}$ is the set of ordered pairs (o, α) for all non-zero integers α.

b. (of o/α): o/α is $\boldsymbol{0}$ for each non-zero integer α.

c. (of *rational number*): A *rational number* is either a dyad α/β or the element $\boldsymbol{0}$.

d. (of $\boldsymbol{0}\cdot\boldsymbol{x}$ and $\boldsymbol{x}\cdot\boldsymbol{0}$): $\boldsymbol{0}\cdot\boldsymbol{x} = \boldsymbol{x}\cdot\boldsymbol{0} = \boldsymbol{0}$ for every rational number $\boldsymbol{x}$.

Note: $\boldsymbol{0}\cdot\boldsymbol{0}$ is twice defined, but is $\boldsymbol{0}$ under each definition, so that there is no inconsistency.

Notation: Bold italic letters will denote rational numbers.

2. Theorem: *If $\beta \neq o$ and $\delta \neq o$, then $\alpha/\beta = \gamma/\delta$ if and only if $\alpha\cdot\delta = \beta\cdot\gamma$.*

Proof: This follows from **B4** if $\alpha \neq o$ and $\gamma \neq o$. If $\alpha = o$, then $\alpha\cdot\delta = o$, by **C21**, and $\alpha/\beta = \boldsymbol{0}$, by **E1b**. But $\gamma/\delta = \boldsymbol{0}$ if and only if $\gamma = o$, by **E1b**; and $\beta\cdot\gamma = o$ if and only if $\gamma = o$, by **C21**.
The proof is similar if $\gamma = o$.

3. Theorem: *If $\beta \neq o$ and $\delta \neq o$, then $(\alpha/\beta)\cdot(\gamma/\delta) = (\alpha\cdot\gamma)/(\beta\cdot\delta)$.*

Proof: This follows from **B7** if $\alpha \neq o$ and $\gamma \neq o$.

If $\alpha = o$, then
$$\begin{aligned} (\alpha/\beta)\cdot(\gamma/\delta) &= \boldsymbol{0}\cdot(\gamma/\delta) && \text{by } \mathbf{E1b} \\ &= \boldsymbol{0} && \text{by } \mathbf{E1d} \\ &= o/(\beta\cdot\delta) && \text{by } \mathbf{E1b} \\ &= (\alpha\cdot\gamma)/(\beta\cdot\delta) && \text{by } \mathbf{C21}. \end{aligned}$$
Similarly if $\gamma = o$.

4. Theorem: *If $\gamma \neq o$ and $\alpha \neq o$, then $\beta/\gamma = (\alpha\cdot\beta)/(\alpha\cdot\gamma)$.*

Proof: This follows from **B8** if $\beta \neq o$. If $\beta = o$, then
$$\begin{aligned} \beta/\gamma &= \boldsymbol{0} && \text{by } \mathbf{E1b} \\ &= o/(\alpha\cdot\gamma) && \text{by } \mathbf{E1b} \\ &= (\alpha\cdot\beta)/(\alpha\cdot\gamma) && \text{by } \mathbf{C21}. \end{aligned}$$

5. Theorem: $\boldsymbol{x}\cdot\boldsymbol{y} = \boldsymbol{y}\cdot\boldsymbol{x}$.

Proof: This follows from **B9** if $\boldsymbol{x} \neq \boldsymbol{0}$ and $\boldsymbol{y} \neq \boldsymbol{0}$; from **E1d** if $\boldsymbol{x} = \boldsymbol{0}$ or $\boldsymbol{y} = \boldsymbol{0}$.

6. Theorem: $(\boldsymbol{x}\cdot\boldsymbol{y})\cdot\boldsymbol{z} = \boldsymbol{x}\cdot(\boldsymbol{y}\cdot\boldsymbol{z})$.

Proof: This follows from **B10** if $\boldsymbol{x} \neq \boldsymbol{0}$, $\boldsymbol{y} \neq \boldsymbol{0}$, and $\boldsymbol{z} \neq \boldsymbol{0}$; from **E1d** if $\boldsymbol{x} = \boldsymbol{0}$ or $\boldsymbol{y} = \boldsymbol{0}$ or $\boldsymbol{z} = \boldsymbol{0}$.

7. Theorem: $\boldsymbol{1}\cdot\boldsymbol{x} = \boldsymbol{x}$.

Proof: This follows from **B14** if $\boldsymbol{x} \neq \boldsymbol{0}$; from **E1d** if $\boldsymbol{x} = \boldsymbol{0}$.

8. Theorem: *If $\boldsymbol{x} \neq \boldsymbol{0}$, then $\boldsymbol{x}\cdot\boldsymbol{x}^{-} = \boldsymbol{1}$.*

Proof: This follows from **B15**.

9. Lemma (to definition **E10**): *If $\beta \neq o$, $\kappa \neq o$, $\delta \neq o$, and $\sigma \neq o$, and*

(i) $$\alpha/\beta = \pi/\kappa$$

and

(ii) $$\gamma/\delta = \rho/\sigma,$$

then

(iii) $$(\alpha\cdot\delta + \beta\cdot\gamma)/\beta\cdot\delta = (\pi\cdot\sigma + \kappa\cdot\rho)/\kappa\cdot\sigma.$$

Proof: By **E2**, (i) and (ii) are true if and only if

(iv) $$\alpha\cdot\kappa = \beta\cdot\pi \quad \text{and} \quad \gamma\cdot\sigma = \delta\cdot\rho.$$

Then $(\alpha\cdot\delta + \beta\cdot\gamma)\cdot(\kappa\cdot\sigma) = (\alpha\cdot\delta)\cdot(\kappa\cdot\sigma) + (\beta\cdot\gamma)\cdot(\kappa\cdot\sigma)$ by **C18** and **20**

$= (\alpha\cdot\kappa)\cdot(\delta\cdot\sigma) + (\gamma\cdot\sigma)\cdot(\beta\cdot\kappa)$ by **C18** and **19**

$= (\beta\cdot\pi)\cdot(\delta\cdot\sigma) + (\delta\cdot\rho)\cdot(\beta\cdot\kappa)$ by (iv)

$= (\beta\cdot\delta)\cdot(\pi\cdot\sigma) + (\beta\cdot\delta)\cdot(\kappa\cdot\rho)$ by **C18** and **19**

$= (\beta\cdot\delta)\cdot(\pi\cdot\sigma + \kappa\cdot\rho)$ by **C20**.

By **C21**, $\kappa\cdot\sigma \neq o$ and $\beta\cdot\delta \neq o$. Therefore (iii) is true, by **E2**.

10. Definition (of *sum* of rational numbers): $\boldsymbol{x}+\boldsymbol{y}$ is $(\alpha\cdot\delta+\beta\cdot\gamma)/\beta\cdot\delta$, where $\alpha/\beta = \boldsymbol{x}$ and $\gamma/\delta = \boldsymbol{y}$.

Note: That this is the same no matter which α, β, γ, and δ are chosen, as long as $\alpha/\beta = \boldsymbol{x}$ and $\gamma/\delta = \boldsymbol{y}$, follows from **E9**. α and γ are integers, and β and δ are integers not equal to o. Therefore $\alpha\cdot\delta+\beta\cdot\gamma$ is an integer, and $\beta\cdot\delta$ is an integer not equal to o, by **C21**. Therefore the sum of two rational numbers is a rational number.

11. Lemma (to various theorems): *If* $\beta \neq 0$, *then*

$$(\alpha/\beta) + (\gamma/\beta) = (\alpha + \gamma)/\beta.$$

Proof: $(\alpha/\beta) + (\gamma/\beta) = (\alpha\cdot\beta + \beta\cdot\gamma)/\beta\cdot\beta$ by definition **E10**

$= (\beta\cdot(\alpha + \gamma))/\beta\cdot\beta$ by **C18** and **20**

$= (\alpha + \gamma)/\beta$ by **E4**.

12. Theorem: $\boldsymbol{x} + \boldsymbol{y} = \boldsymbol{y} + \boldsymbol{x}$.

Proof: Let $\boldsymbol{x}$ be α/β and $\boldsymbol{y}$ be γ/δ.

Then $\boldsymbol{x} + \boldsymbol{y} = (\alpha\cdot\delta + \beta\cdot\gamma)/\beta\cdot\delta$ by definition **E10**

$= (\delta\cdot\alpha + \gamma\cdot\beta)/\delta\cdot\beta$ by **C18**

$= (\gamma\cdot\beta + \delta\cdot\alpha)/\delta\cdot\beta$ by **C6**

$= (\gamma/\delta) + (\alpha/\beta)$ by definition **E10**

$= \boldsymbol{y} + \boldsymbol{x}$.

13. Theorem: $(\boldsymbol{x} + \boldsymbol{y}) + \boldsymbol{z} = \boldsymbol{x} + (\boldsymbol{y} + \boldsymbol{z})$.

Proof: Let $\boldsymbol{x}$ be α/β, $\boldsymbol{y}$ be γ/δ, and $\boldsymbol{z}$ be θ/ζ. Then

$(\boldsymbol{x} + \boldsymbol{y}) + \boldsymbol{z} = ((\alpha\cdot\delta + \beta\cdot\gamma)/\beta\cdot\delta) + (\theta/\zeta)$ by definition **E10**

$= ((\alpha\cdot\delta + \beta\cdot\gamma)\cdot\zeta + (\beta\cdot\delta)\cdot\theta)/(\beta\cdot\delta)\cdot\zeta$ by definition **E10**

$= (\alpha\cdot(\delta\cdot\zeta) + \beta\cdot(\gamma\cdot\zeta + \delta\cdot\theta))/\beta\cdot(\delta\cdot\zeta)$ by **C18, 19**, and **20**

$= (\alpha/\beta) + ((\gamma\cdot\zeta + \delta\cdot\theta)/\delta\cdot\zeta)$ by definition **E10**

$= (\alpha/\beta) + ((\gamma/\delta) + (\theta/\zeta))$ by definition **E10**

$= \boldsymbol{x} + (\boldsymbol{y} + \boldsymbol{z})$.

14. Theorem: *For each* $\mathbf{a}$ *there is a* $\mathbf{x}$ *for which* $\mathbf{a} + \mathbf{x} = \mathbf{0}$.

Proof: If $\mathbf{a}$ is α/β, let $\mathbf{x}$ be $(-\alpha)/\beta$. Then

$$
\begin{aligned}
\mathbf{a} + \mathbf{x} &= (\alpha + (-\alpha))/\beta && \text{by } \mathbf{E11} \\
&= o/\beta && \text{by } \mathbf{C11} \\
&= \mathbf{0} && \text{by definition } \mathbf{E1b}.
\end{aligned}
$$

15. Theorem: $\mathbf{0} + \mathbf{x} = \mathbf{x}$.

Proof: Let $\mathbf{x}$ be α/β. $\mathbf{0} = o/\beta$ by definition **E1b**.

Then

$$
\begin{aligned}
\mathbf{0} + \mathbf{x} &= (o + \alpha)/\beta && \text{by } \mathbf{E11} \\
&= \alpha/\beta && \text{by } \mathbf{C10} \\
&= \mathbf{x}.
\end{aligned}
$$

16. Theorem: $\mathbf{x}\cdot(\mathbf{y} + \mathbf{z}) = \mathbf{x}\cdot\mathbf{y} + \mathbf{x}\cdot\mathbf{z}$.

Proof: Let $\mathbf{x}$ be α/β, $\mathbf{y}$ be γ/δ, and $\mathbf{z}$ be θ/ζ. Then

$$
\begin{aligned}
\mathbf{x}\cdot(\mathbf{y} + \mathbf{z}) &= (\alpha/\beta)\cdot((\gamma\cdot\zeta + \delta\cdot\theta)/\delta\cdot\zeta) && \text{by definition } \mathbf{E10} \\
&= (\alpha\cdot(\gamma\cdot\zeta + \delta\cdot\theta))/\beta\cdot(\delta\cdot\zeta) && \text{by } \mathbf{E3} \\
&= (\beta\cdot(\alpha\cdot(\gamma\cdot\zeta + \delta\cdot\theta)))/\beta\cdot(\beta\cdot(\delta\cdot\zeta)) && \text{by } \mathbf{E4} \\
&= ((\alpha\cdot\gamma)\cdot(\beta\cdot\zeta) + (\beta\cdot\delta)\cdot(\alpha\cdot\theta))/(\beta\cdot\delta)\cdot(\beta\cdot\zeta) && \text{by } \mathbf{C18, 19,} \text{ and } \mathbf{20} \\
&= (\alpha\cdot\gamma/\beta\cdot\delta) + (\alpha\cdot\theta/\beta\cdot\zeta) && \text{by definition } \mathbf{E10} \\
&= (\alpha/\beta)\cdot(\gamma/\delta) + (\alpha/\beta)\cdot(\theta/\zeta) && \text{by } \mathbf{E3} \\
&= \mathbf{x}\cdot\mathbf{y} + \mathbf{x}\cdot\mathbf{z}.
\end{aligned}
$$

17. Theorem: *The correspondence* $\xi/o' \longleftrightarrow \xi$ *between rational numbers of the form* ξ/o' *and the integers is a multiplication- and addition-isomorphism.*

Proof: By **B11**, the correspondence is a matching between rational numbers of the form ξ/o' where $\xi \neq o$ and integers other than o. There is, by definition **E1**, only one number of the form o/o'; and this corresponds to o. Therefore the complete correspondence is a matching. By **B11** and **E1d**, it is a multiplication-isomorphism.

$$(\xi/o') + (\eta/o') = (\xi + \eta)/o' \quad \text{by } \mathbf{E11}.$$

Therefore the correspondence is an addition-isomorphism.

Definition (of *integral* rational number): A rational number is *integral* if it corresponds to an integer in this isomorphism; that is, if it is of the form ξ/o'.

18. Theorem: $\boldsymbol{0} \neq \boldsymbol{1}$.

Proof: $\boldsymbol{0} = o/o'$, by definition **E1b**, and $\boldsymbol{1} = o'/o'$, by notation **E1**. Therefore they correspond to o and o' respectively in isomorphism **E17**. These correspond to 0 and $0'$ in isomorphism **C23**. $0 \neq 0'$, by **A1b**. Therefore $o \neq o'$, and so $\boldsymbol{0} \neq \boldsymbol{1}$.

19. Theorem: *The set of rational numbers with the operations* $+$ *and* $\cdot$ *form a field.*

Proof: **D1a-k** follow from **E13, 12, 15, 14, 6, 5, 7** and **18, 8,** and **16** respectively.

Note: We have now proved that all the results of **D2-9** are true for rational numbers.

Further, if $\beta \neq o$, then

$$-(\alpha/\beta) = (-\alpha)/\beta \qquad \text{by } \textbf{D2c} \text{ and the proof of } \textbf{E14}$$

and if also $\alpha \neq o$, then

$$(\alpha/\beta)^- = \beta/\alpha \qquad \text{by } \textbf{B15} \text{ and } \textbf{D5c}.$$

20. Lemma (to **G2**): If $\beta \neq o$, then $(-\alpha)/\beta = \alpha/(-\beta)$.

Proof: $(-\alpha)\cdot(-\beta) = \beta\cdot\alpha$ by **C29** and **C18**.

Therefore $(-\alpha)/\beta = \alpha/(-\beta)$ by **E2**.

EXERCISES E

1. Prove that if $\boldsymbol{x}$ is any rational number, there is a rational number $\boldsymbol{y}$ such that $\boldsymbol{y} + (\boldsymbol{y} + \boldsymbol{y}) = \boldsymbol{x}$.

2*. Prove that there is no rational number $\boldsymbol{x}$ for which $\boldsymbol{x}\cdot\boldsymbol{x} = \boldsymbol{2}$.

3*. Prove that if α and β are positive non-zero integers with no common factor, there is a rational number $\boldsymbol{x}$ for which $\boldsymbol{x}\cdot\boldsymbol{x} = \alpha/\beta$ if and only if there are rational numbers $\boldsymbol{y}$ and $\boldsymbol{z}$ for which $\boldsymbol{y}\cdot\boldsymbol{y} = \alpha/o'$ and $\boldsymbol{z}\cdot\boldsymbol{z} = \beta/o'$.

4. T is the set of all ordered pairs of rational numbers. Addition and multiplication are defined in T by the rules:

$$(\boldsymbol{a}, \boldsymbol{b}) + (\boldsymbol{c}, \boldsymbol{d}) = (\boldsymbol{a} + \boldsymbol{c}, \boldsymbol{b} + \boldsymbol{d})$$
$$(\boldsymbol{a}, \boldsymbol{b})\cdot(\boldsymbol{c}, \boldsymbol{d}) = (\boldsymbol{a}\cdot\boldsymbol{c} + \boldsymbol{2}\cdot(\boldsymbol{b}\cdot\boldsymbol{d}), \boldsymbol{a}\cdot\boldsymbol{d} + \boldsymbol{b}\cdot\boldsymbol{c}).$$

Prove that T is a field and that $(\boldsymbol{x}, \boldsymbol{0}) \longleftrightarrow \boldsymbol{x}$ is an isomorphism between certain elements of T and the rational numbers.

Prove that T contains an element ξ for which $\xi\cdot\xi = (\boldsymbol{2}, \boldsymbol{0})$.

CHAPTER **F**

Ordered Fields

1. Definition (of *ordered field*): An *ordered field* is a field some of whose elements form a set P with the following properties.

a. If x is any element of the field, then either $x \in P$ or $-x \in P$.

b. If both $x \in P$ and $-x \in P$, then $x = o$.

c. If $x \in P$ and $y \in P$, then $x + y \in P$ and $x \cdot y \in P$.

Notation: In Chapter **F**, small italic letters will denote elements of an ordered field. (o and i will be, as in Chapter **D**, elements such that $o + x = x$ and $i \cdot x = x$ for all x.)

2. Theorem: $o \in P$.

Proof: $o \in P$ or $-o \in P$ by **F1a.**

But $o = -o$ by **D2e.**

3. Theorem: $x \cdot x \in P$.

Proof: If $x \in P$, then $x \cdot x \in P$ by **F1c.**

If $x \notin P$, then $-x \in P$ by **F1a.**

Then $x \cdot x = (-x) \cdot (-x)$ by **D9**

$\in P$ by **F1c.**

4. Theorem: $i \in P$ *and* $-i \notin P$.

Proof: $i = i \cdot i$ by **D1g**

$\in P$ by **F3.**

If now $-i \in P$, we should have $i = o$, by **F1b**. But this is not so, by **D1g.**

5. Theorem: *If* $x \in P$ *and* $x \neq o$, *then* $x^- \in P$.

Proof: If $x^- \notin P$, then $-x^- \in P$ by **F1a.**

Therefore $-i = -(x \cdot x^-)$ by **D1h**

$= x \cdot (-x^-)$ by **D8**

$\in P$ by **F1c**

which is not so, by **F4.**

6. Definition (of $\geqslant$): If $x + (-y) \in P$, then $x \geqslant y$.

7. Theorem: *$x \geqslant o$ if and only if $x \in P$.*

Proof: By definition **F6**, $x \geqslant o$ if and only if

$$x + (-o) \in P.$$

But $x + (-o) = x + o$ by **D2e**

$= x$ by **D1b** and **c**.

8. Theorem: *Either $x \geqslant y$ or $y \geqslant x$.*

Proof: Either (I) $x + (-y) \in P$ or (II) $-(x + (-y)) \in P$, by **F1a**. (I) implies $x \geqslant y$, by definition **F6**.

Now $-(x + (-y)) = -x + (-(-y))$ by **D2f**

$= -x + y$ by **D2d**

$= y + (-x)$ by **D1b**.

Therefore (II) implies $y \geqslant x$, by definition **F6**.

9. Theorem: *$x \geqslant x$.*

Proof: $x + (-x) = o$ by **D1d**

$\in P$ by **F2**.

Therefore $x \geqslant x$ by definition **F6**.

10. Theorem: *If $x \geqslant y$ and $y \geqslant z$, then $x \geqslant z$.*

Proof: If $x \geqslant y$, then $x + (-y) \in P$ by definition **F6**.

If $y \geqslant z$, then $y + (-z) \in P$ by definition **F6**.

Therefore, by **F1c**, $(x + (-y)) + (y + (-z)) \in P$

$= (x + (y + (-y))) + (-z)$ by **D1a** and **b**

$= (x + o) + (-z)$ by **D1d**

$= x + (-z)$ by **D1b** and **c**.

Therefore $x \geqslant z$ by definition **F6**.

Notation: We shall write $x \geqslant y \geqslant z$ in place of

$x \geqslant y$ and $y \geqslant z$.

11. Theorem: *If $x_1 \geqslant x_2$ and $y_1 \geqslant y_2$, then $x_1 + y_1 \geqslant x_2 + y_2$.*

Proof: If $x_1 \geqslant x_2$, then $x_1 + (-x_2) \in P$ by definition **F6**.

If $y_1 \geqslant y_2$, then $y_1 + (-y_2) \in P$ by definition **F6**.

Therefore	$(x_1 + (-x_2)) + (y_1 + (-y_2)) \in P$	by **F1c**
	$= (x_1 + y_1) + (-x_2 + (-y_2))$	by **D1a** and **b**
	$= (x_1 + y_1) + (-(x_2 + y_2))$	by **D2f.**
Therefore	$x_1 + y_1 \geqslant x_2 + y_2$	by definition **F6.**

12. Lemma (to **F13**): *If $y \in P$ and if $x_1 \geqslant x_2$, then $x_1 \cdot y \geqslant x_2 \cdot y$.*

Proof:	$x_1 + (-x_2) \in P$	by **F6.**
Therefore	$(x_1 + (-x_2)) \cdot y \in P$	by **F1c**
	$= x_1 \cdot y + (-(x_2 \cdot y))$	by **D7** and **8.**
Therefore	$x_1 \cdot y \geqslant x_2 \cdot y$	by definition **F6.**

13. Theorem: *If x_2 and y_1 are in P, and if $x_1 \geqslant x_2$ and $y_1 \geqslant y_2$, then $x_1 \cdot y_1 \geqslant x_2 \cdot y_2$.*

Proof:	$x_1 \cdot y_1 \geqslant x_2 \cdot y_1$	by **F12**
	$= y_1 \cdot x_2$	by **D1f**
	$\geqslant y_2 \cdot x_2$	by **F12**
	$= x_2 \cdot y_2$	by **D1f.**
Therefore	$x_1 \cdot y_1 \geqslant x_2 \cdot y_2$	by **F10.**

14. Theorem: *If $x \in P$, $y \in P$, $x \neq o$, $y \neq o$, and $x \geqslant y$, then $y^- \geqslant x^-$.*

Proof: If $x \geqslant y$, then $x + (-y) \in P$, by definition **F6**. y^- and x^- are in P, by **F5**.

Therefore	$(x^- \cdot (x + (-y))) \cdot y^- \in P$	by **F1c**
	$= (x^- \cdot x + x^- \cdot (-y)) \cdot y^-$	by **D1k**
	$= (i + (-x^-) \cdot y) \cdot y^-$	by **D1f**, **1h**, and **8**
	$= i \cdot y^- + (-x^-) \cdot (y \cdot y^-)$	by **D7** and **D1e**
	$= y^- + (-x^-)$	by **D1g**, **h**, and **f.**
Therefore	$y^- \geqslant x^-$	by definition **F6.**

15. Definition (of $|\ |$): $\begin{cases} |x| = x \text{ if } x \in P \\ |x| = -x \text{ if } x \notin P. \end{cases}$

16. Notes: If $-x \in P$, then, by **F1a** and **b**, either $x = o$ or $x \notin P$; in either case, $|x| = -x$.

Whatever x is, $|x| = |-x| \in P$.

17. Theorem: $|x|\cdot|y| = |x\cdot y|$.

Proof: By definition **F15**, $|x|\cdot|y|$ is either $x\cdot y$, $x\cdot(-y)$, $(-x)\cdot y$, or $(-x)\cdot(-y)$. Therefore, by **D8** and **9**, $|x|\cdot|y|$ is either $x\cdot y$ or $-(x\cdot y)$. Therefore, by note **F16**, $||x|\cdot|y|| = |x\cdot y|$.

But $|x|\cdot|y| \in P$, by **F1c** and note **F16**. Therefore $||x|\cdot|y|| = |x|\cdot|y|$, by definition **F15**.

18. Theorem: *If* $x \neq o$, *then* $|x^-| = |x|^-$.

Proof:	$|x|\cdot|x^-| = |x\cdot x^-|$	by **F17**
	$= |i|$	by **D1h**
	$= i$	by **F4** and definition **F15**.
Therefore	$|x^-| = |x|^-$	by **D5b**.

19. Theorem: $|x| \geqslant x$.

Proof:	If $x \in P$, then $|x| = x$	by definition **F15**
	$\geqslant x$	by **F9**.
If	$x \notin P$, then $|x| = -x$	by definition **F15**
and	$-x \in P$	by **F1a**.
Then	$-x + (-x) \in P$	by **F1c**
that is	$-x \geqslant x$	by definition **F6**.

20. Theorem: $|x| + |y| \geqslant |x + y|$.

Proof:	$|x| \geqslant x$ and $|y| \geqslant y$	by **F19**.
Therefore	$|x| + |y| \geqslant x + y$	by **F11**.
Again	$|x| + |y| = |-x| + |-y|$	by note **F16**
	$\geqslant -x + (-y)$	as above
	$= -(x + y)$	by **D2f**.
But	$|x + y|$ is either $x + y$ or $-(x + y)$	by definition **F15**.
Therefore	$|x| + |y| \geqslant |x + y|$.	

21. Theorem: *The equations*

(i) $$0^\times = o$$

and

(ii) $$(x')^\times = i + x^\times$$

define an element $x^\times$ *of an ordered field for each whole number* x. *The set of all elements* $x^\times$ *is isomorphic to the set of all whole numbers.*

Proof: Let M be the set of all whole numbers x for which $x^\times$ is defined. $0 \in M$ by (i), and if $x \in M$ then $x' \in M$ by (ii). Therefore every whole number is in M by **A1c**.

$0^\times \in P$, by (i) and **F2**. If $x^\times \in P$, then $(x')^\times \in P$ by (ii), **F4**, and **F1c**. Therefore, by **A1c**, every $x^\times \in P$.

If $(x')^\times = o$, then $i + x^\times = o$, by (ii), and so $x^\times = -i$, by **D2c**, which is impossible because every $x^\times \in P$ whereas, by **F4**, $-i \notin P$. Therefore, for every x, $(x')^\times \neq o$.

Now let N be the set of all y for which $x^\times = y^\times$ only if $x = y$. If $x \neq 0$, then $x = u'$ for some u by **A2** and so $x^\times = (u')^\times \neq o$. Therefore $0 \in N$. If $y \in N$, suppose that $x^\times = (y')^\times$. Then $x^\times \neq o$ and so $x \neq 0$, because $0 \in N$. Then $x = u'$ for some u by **A2**. Then, by (ii), $i + u^\times = (u')^\times = (x)^\times = (y')^\times = i + y^\times$. Therefore $u^\times = y^\times$, by **D2a**, and so $u = y$, because $y \in N$. Then $x = u' = y'$. Therefore $y' \in N$ and so, by **A1c**, every y is in N. Therefore $x^\times = y^\times$ if and only if $x = y$, and so $x \longleftrightarrow x^\times$ is a matching.

Let L be the set of y for which $(x + y)^\times = x^\times + y^\times$ for every x.

$0 \in L$, because
$$\begin{aligned}(x + 0)^\times &= x^\times && \text{by } \mathbf{A3a}\\ &= x^\times + o && \text{by } \mathbf{D1b} \text{ and } \mathbf{c}\\ &= x^\times + 0^\times && \text{by (i).}\end{aligned}$$

If $y \in L$, then
$$\begin{aligned}(x + y')^\times &= ((x + y)')^\times && \text{by } \mathbf{A3b}\\ &= i + (x + y)^\times && \text{by (ii)}\\ &= i + (x^\times + y^\times) && \text{because } y \in L\\ &= x^\times + (i + y^\times) && \text{by } \mathbf{D1a} \text{ and } \mathbf{b}\\ &= x^\times + (y')^\times && \text{by (ii).}\end{aligned}$$

Therefore $y' \in L$ and so, by **A1c**, every y is in L. Therefore $(x + y)^\times = x^\times + y^\times$ for every x and y. Similarly, using **A9** in place of **A3**, we prove that $(x \cdot y)^\times = x^\times \cdot y^\times$ for every x and y. Therefore $x \longleftrightarrow x^\times$ is an isomorphism.

Definition (of *integer* of an ordered field): The elements $x^\times$ and $-x^\times$ are the *integers* of the ordered field.

22. Lemma (to **F23**): $i + i \neq o$.

Proof: If $i + i = o$, then $i = -i$, by **D2c**. But $i \in P$ and $-i \notin P$, by **F4**.

23. Notation: By **D6** and **F22**, there is, given b, just one x for which $(i + i)\cdot x = b$. We denote it by $\frac{1}{2}b$. Then

$$\begin{aligned} \tfrac{1}{2}b + \tfrac{1}{2}b &= x + x \\ &= i\cdot x + i\cdot x && \text{by } \mathbf{D1g} \\ &= (i + i)\cdot x && \text{by } \mathbf{D1k} \text{ and } \mathbf{f} \\ &= b. \end{aligned}$$

24. Theorem: *If* $b \neq o$, *then* $\frac{1}{2}b \neq o$. *If also* $b \in P$, *then* $\frac{1}{2}b \in P$.

Proof: $b = (i + i)\cdot\frac{1}{2}b$, and $i + i \neq o$, by **F22**. Therefore $b = o$ if and only if $\frac{1}{2}b = o$, by **D3** and **4**.

Now suppose that $b \neq o$ and $b \in P$.

If $\quad \frac{1}{2}b \notin P$, then $-\frac{1}{2}b \in P \quad$ by **F1a.**

Then $\quad -b = -(\frac{1}{2}b + \frac{1}{2}b) = -\frac{1}{2}b + (-\frac{1}{2}b) \quad$ by **D2f**

$\in P \quad$ by **F1c.**

This is impossible by **F1b**, and so $\frac{1}{2}b \in P$.

EXERCISES F

1. Prove that if $x \geqslant y$, then $-y \geqslant -x$.
2. Prove that if $o \geqslant x \geqslant y$ and $x \neq o$, then $y \neq o$ and $y^- \geqslant x^-$.
3. Prove that if $a \in P$ then $i + a \neq o$.
4. Prove that if we define x' to be $x + i$, and I to be the least set which contains o and contains x' whenever it contains x, then I is a succession. Prove also that addition and multiplication defined for this succession as in **A3** and **9** are the same as addition and multiplication in the ordered field. Are the results still true if we take our elements from a field which is not ordered?
5. Show how to make the field T of example **E4** an ordered field.

CHAPTER G

The Order-Relation for Rational Numbers

1. Definition (of *positive* rational number): If α and β are positive and β is not o, then α/β is *positive.*

Note: o and o' are positive, by note **C25**. $\boldsymbol{0} = o/o'$ and $\boldsymbol{1} = o'/o'$. Therefore $\boldsymbol{0}$ and $\boldsymbol{1}$ are positive.

2. Theorem: *Either* $\mathbf{x}$ *or* $-\mathbf{x}$ *is positive.*

Proof: Let $\mathbf{x}$ be α/β. By **C27**, either α or α is positive, and either β or $-\beta$ is positive. β is not o, by definition **E1c**.

If α and β are positive, then α/β is positive, by definition **G1**.

If $-\alpha$ and β are positive, then $-\mathbf{x} = (-\alpha)/\beta$ by **E19**
which is positive, by definition **G1**.

If α and $-\beta$ are positive, then $-\mathbf{x} = (-\alpha)/\beta$ by **E19**
$= \alpha/(-\beta)$ by **E20**
which is positive, by definition **G1**.

If $-\alpha$ and $-\beta$ are positive, then $\mathbf{x} = (-\alpha)/(-\beta)$, by **E20** and **D2d**
which is positive, by definition **G1**.

3. Theorem: *If* $\mathbf{x}$ *and* $-\mathbf{x}$ *are both positive, then* $\mathbf{x} = \boldsymbol{0}$.

Proof: If $\mathbf{x}$ is positive, then $\mathbf{x} = \alpha/\beta$ where α and β are positive, by definition **G1**. Then

$$-\mathbf{x} = (-\alpha)/\beta \qquad \text{by } \mathbf{E19}.$$

If $-\mathbf{x}$ is positive, then $-\mathbf{x} = \gamma/\delta$

where γ and δ are positive and δ is non-zero, by definition **G1**. Therefore

$$\begin{aligned} \beta\cdot\gamma &= (-\alpha)\cdot\delta && \text{by } \mathbf{E2} \\ &= -(\alpha\cdot\delta) && \text{by } \mathbf{C28}. \end{aligned}$$

$\beta\cdot\gamma$ and $\alpha\cdot\delta$ are positive, by **C30**.

Therefore $\alpha \cdot \delta$ and $-\alpha \cdot \delta$ are both positive.
Therefore $\alpha \cdot \delta = o$, by **C31**.
Therefore $\alpha = o$, by **C21** and because $\delta \neq o$.
Therefore $\boldsymbol{x} = o/\beta = \boldsymbol{0}$, by definition **E1b**.

4. Theorem: *If* $\boldsymbol{x}$ *and* $\boldsymbol{y}$ *are positive, then so are* $\boldsymbol{x} + \boldsymbol{y}$ *and* $\boldsymbol{x} \cdot \boldsymbol{y}$.

Proof: Let $\boldsymbol{x}$ be α/β and $\boldsymbol{y}$ be γ/δ where α, β, γ, and δ are positive (definition **G1**).

Then $\boldsymbol{x} + \boldsymbol{y} = (\alpha \cdot \delta + \beta \cdot \gamma)/\beta \cdot \delta$ by definition **E10**
which is positive, by **C30** and definition **G1**.

And $\boldsymbol{x} \cdot \boldsymbol{y} = \alpha \cdot \gamma / \beta \cdot \delta$ by **E3**
which is positive, by **C30** and definition **G1**.

5. Theorem: *The rational numbers form an ordered field.*

Proof: Let F be the set of rational numbers and P the set of positive rational numbers. Then **F1a**, **b**, and **c** follow from **G2**, **3**, and **4**.

Note: We may now apply all the definitions and theorems of Chapter **F** to the rational numbers. For example, if $\boldsymbol{x}$ and $\boldsymbol{y}$ are rational numbers, we write $\boldsymbol{x} \geqslant \boldsymbol{y}$ if and only if $\boldsymbol{x} + (-\boldsymbol{y})$ is positive.

6. Theorem: *An integral rational number is positive if and only if it corresponds to a positive integer in isomorphism* **E17**.

Proof: An integral rational number $\boldsymbol{x}$ is of the form ξ/o', by definition **E17**, and corresponds to ξ. o' is positive, by **C25**. Therefore if ξ is positive, so is $\boldsymbol{x}$, by definition **G1**.

On the other hand, if $\boldsymbol{x}$ is positive, then, by definition **G1**, it is equal to α/β for some positive α and β. $\beta \neq o$, by **E1c**. Then $\xi \cdot \beta = o' \cdot \alpha$, by **E2**, which is positive, by **C30**. Therefore ξ is positive, by **C32**.

7. From now on, we shall have no more to do with whole numbers or integers, but shall use in their place the positive integral rational numbers and the integral rational numbers respectively. We can, with this understanding, now abbreviate the phrase " integral rational number " to " integer ". Similarly, we shall refer to the integers of an ordered field (**F21**) simply as " integers ".

8. Lemma (to **H9** and **J2**): *If $\boldsymbol{k}$ is any rational number, there is an integer $\boldsymbol{n}$ such that $\boldsymbol{n} \geqslant \boldsymbol{k}$.*

Proof: If $\boldsymbol{k}$ is positive, let $\boldsymbol{k}$ be α/β, where α and β are positive (definition **G1**). Let $\boldsymbol{n}$ be the integer α/o'.

Then $\boldsymbol{n} + (-\boldsymbol{k}) = (\alpha{\cdot}\beta + o'{\cdot}(-\alpha))/o'{\cdot}\beta$ by **E10** and **19**

$= (\alpha{\cdot}\beta + (-\alpha))/\beta$ by **C26**.

But $\alpha{\cdot}\beta + (-\alpha)$ is positive, by **C33**, and β is positive.

Therefore $\boldsymbol{n} + (-\boldsymbol{k})$ is positive, by definition **G1**, and so $\boldsymbol{n} \geqslant \boldsymbol{k}$, by definition **F6**.

If $\boldsymbol{k}$ is not positive, let $\boldsymbol{n}$ be any positive integer. Then $\boldsymbol{n} \geqslant \boldsymbol{0} \geqslant \boldsymbol{k}$, by **F7** and **8**.

EXERCISES G

1. Prove that if $\boldsymbol{p}$ is a rational number there are integers α and β such that $\beta \geqslant o$ and $\boldsymbol{p} = \alpha/\beta$.
2. Prove that if $\boldsymbol{p}$ and $\boldsymbol{q}$ are rational numbers there are integers α, β, and γ such that $\boldsymbol{p} = \alpha/\gamma$ and $\boldsymbol{q} = \alpha/\gamma$.
3. Prove that if $\boldsymbol{p}$ and $\boldsymbol{q}$ are rational numbers and $\boldsymbol{p} \geqslant \boldsymbol{q}$ but $\boldsymbol{p} \neq \boldsymbol{q}$, then there is a rational number $\boldsymbol{r}$ such that $\boldsymbol{p} \geqslant \boldsymbol{r} \geqslant \boldsymbol{q}$ and $\boldsymbol{p} \neq \boldsymbol{r} \neq \boldsymbol{q}$.
4. Prove that if $\boldsymbol{x}$ is a rational number, there are positive rational numbers $\boldsymbol{y}$ and $\boldsymbol{z}$ such that $\boldsymbol{x} = \boldsymbol{y} + (-\boldsymbol{z})$.
5. Prove that if $\boldsymbol{M}$ is a non-null set of positive integers, then there is an element $\boldsymbol{k}$ in $\boldsymbol{M}$ such that $\boldsymbol{m} \geqslant \boldsymbol{k}$ whenever $\boldsymbol{m} \in \boldsymbol{M}$.
6. Prove that if $\boldsymbol{x}$ is a non-integral rational number, then there is one and only one integral rational number $\boldsymbol{p}$ such that $\boldsymbol{p} + \boldsymbol{1} \geqslant \boldsymbol{x} \geqslant \boldsymbol{p}$.
7. Prove that if $\boldsymbol{p}$ and $\boldsymbol{q}$ are positive integers, then $\boldsymbol{q} \geqslant \boldsymbol{p}$ if and only if there is a positive integer $\boldsymbol{r}$ such that $\boldsymbol{p} + \boldsymbol{r} = \boldsymbol{q}$.

CHAPTER H

Exponentiation

1. Notation: In **H2–8**, small italic letters other than p, q, r will denote elements of a field F; in **H10–13** they will denote elements of an ordered field. We shall write

$$x - y \quad \text{for} \quad x + (-y)$$

and

$$x/y \quad \text{for} \quad x \cdot y^{-}$$

$x - y$ is defined for all x and y of F. Therefore we may look on $-$ as an operation; we call it *subtraction*. x/y is defined for all x and all non-zero y of F. / is often looked on as an operation, named *division*, with the proviso that *division by zero is not possible.* Theorems involving these two operations can be proved straight from their definitions and the properties of addition and multiplication. For example:

$$(x - y) + (y - z) = x - z.$$

Proof:

$$\begin{aligned} (x - y) + (y - z) &= (x + (-y)) + (y + (-z)) \\ &= x + ((-y) + (y + (-z))) && \text{by } \textbf{D1a} \\ &= x + (((-y) + y) + (-z)) && \text{by } \textbf{D1a} \\ &= x + (0 + (-z)) && \text{by } \textbf{D1d} \text{ and } \textbf{b} \\ &= x - z && \text{by } \textbf{D1c}. \end{aligned}$$

Other such formulæ are

$$x - x = 0$$

$$(x - y) + (z - w) = (x + z) - (y + w)$$

If $u \neq 0$ *and* $v \neq 0$, *then* $x/u + y/v = (x \cdot v + u \cdot y)/u \cdot v$.

A list of 27 such theorems, with proofs, will be found in E. Landau's *Grundlagen der Analysis*, p. 101. The reader will readily think of others for himself.

Besides the abbreviations $x - y$ and x/y defined above, we shall from this point on use

$$x + y + z \quad \text{for} \quad (x + y) + z \qquad \text{(compare } \mathbf{B2}\text{)}$$
$$x + y - z \quad \text{for} \quad (x + y) - z$$
$$x{\cdot}y{\cdot}z \quad \text{for} \quad (x{\cdot}y){\cdot}z$$

and so on; in fact, we shall use the familiar notation of elementary algebra. We shall no longer painstakingly refer to the laws of arithmetic each time one is used. Our notation even conceals some of these laws: $x + y + z$ will be used freely for $x + (y + z)$, thus concealing the associative law of addition.

If the field is an ordered field we shall write $x \geqslant 0$ instead of $x \in P$ (using **F7**) and shall call the elements of P *positive*. We shall write $x > y$ to mean

$$x \geqslant y \text{ but } x \neq y.$$

We shall also use $x < y$ for $y > x$, and $x \leqslant y$ for $y \geqslant x$. Again, many theorems well known in elementary algebra may be proved straight from the definitions, for example:

If $x > y \geqslant z$, then $x > z$.

We shall use the matching between positive integers and whole numbers to enable us to give proofs by induction (that is, using **A1c**) for the positive integers.

We shall use the familiar summation notation, namely that $\sum_{i=1}^{1} x_i = x_1$, and $\sum_{i=1}^{n+1} x_i = \sum_{i=1}^{n} x_i + x_{n+1}$. (These two equations constitute an inductive definition of $\sum_{i=1}^{n} x_i$.)

2. Theorem: *There is a function x^p, defined for all non-zero x and all integers p, such that*

a. $x^0 = 1$,

b. $x^{p+1} = x^p{\cdot}x$,

and

c. $x^{-p} = (x^-)^p$.

Proof: Given m, we shall define a function g for which $g(n, m)$ is defined whenever $0 \leqslant n \leqslant m$, and

(i) $g(0, m) = 1$,

and

(ii) $g(n + 1, m) = g(n, m){\cdot}x$ whenever $0 \leqslant n \leqslant m - 1$.

First, put $g(0, 0) = 1$. Then, if $m = 0$, (i) and (ii) are true. Thus if

M is the set of all m for which such a function can be defined, then $0 \in M$.

If $m \in M$, then $g(n, m)$ is defined whenever $0 \leqslant n \leqslant m$.

Put $g(n, m+1) = g(n, m)$ whenever $0 \leqslant n \leqslant m$, and put $g(m+1, m+1) = g(m, m) \cdot x$. Then $g(n, m+1)$ is defined whenever $0 \leqslant n \leqslant m+1$.

Now $g(0, m+1) = g(0, m) = 1$. And if $0 \leqslant n \leqslant m-1$, then

$$g(n+1, m+1) = g(n+1, m) = g(n, m) \cdot x = g(n, m+1) \cdot x.$$

Thus $m + 1 \in M$ and so, by induction, every positive integer belongs to M.

Now put

(iii) $\qquad x^p = g(p, p)$ whenever $p \geqslant 0$

and

(iv) $\qquad x^p = (x^-)^{-p}$ whenever $p < 0$.

Then x^p is defined for every integer p; and **c** is satisfied whenever $p > 0$. Putting x^- in place of x in (iv), we have $(x^-)^p = x^{-p}$ whenever $p < 0$; i.e. **c** is satisfied whenever $p < 0$. And finally $x^{-0} = x^0 = 1 = (x^-)^0$; and so **c** is satisfied for every p.

$x^0 = g(0, 0) = 1$; therefore **a** is satisfied.

If $p \geqslant 0$, then $x^{p+1} = g(p+1, p+1) = g(p, p) \cdot x = x^p \cdot x$. Therefore **b** is satisfied if $p \geqslant 0$.

If $p < 0$, then $x^{p+1} \cdot x^- = (x^-)^{-p-1} \cdot x^-$ by **c**

$= (x^-)^{-p}$ by **b**, because $-p - 1 \geqslant 0$

$= x^p$ by **c**.

Therefore $x^{p+1} = x^p \cdot x$. Therefore **b** is satisfied for every integer p.

Definition (of integral *power*): x^p is the pth *power* of x.

3. Theorem: $1^p = 1$.

Proof: $1^0 = 1$, by **H2a**. If $1^p = 1$, then $1^{p+1} = 1^p \cdot 1 = 1 \cdot 1 = 1$. Therefore, by induction, $1^p = 1$ whenever $p \geqslant 0$. If $0 > p$, then $1^p = (1^-)^{-p} = 1^p = 1$.

4. Theorem: $x^p \cdot y^p = (x \cdot y)^p$.

Proof: $x^0 \cdot y^0 = 1 \cdot 1 = 1 = (x \cdot y)^0$.

If $x^p \cdot y^p = (x \cdot y)^p$ then

$$x^{p+1} \cdot y^{p+1} = x^p \cdot x \cdot y^p \cdot y = (x \cdot y)^p x \cdot y = (x \cdot y)^{p+1}.$$

Therefore, by induction, $x^p \cdot y^p = (x \cdot y)^p$ whenever $p \geqslant 0$.

If $0 > p$, then $x^p \cdot y^p = (x^-)^{-p} \cdot (y^-)^{-p}$ by **H2c**

$= (x^- \cdot y^-)^{-p}$ because $-p \geqslant 0$

$= ((x \cdot y)^-)^{-p}$

$= (x \cdot y)^p$ by **H2c.**

5. Theorem: $x^{-p} = (x^p)^-$.

Proof: $x^p \cdot (x^-)^p = (x \cdot x^-)^p$ by **H4**

$= 1^p$

$= 1$ by **H3.**

Therefore $(x^p)^- = (x^-)^p = x^{-p}$.

6. Theorem: $x^p \cdot x^q = x^{p+q}$.

Proof: Let M be the set of q for which $x^p \cdot x^q = x^{p+q}$ for all p. $x^p \cdot x^0 = x^p \cdot 1 = x^p = x^{p+0}$, and so $0 \in M$.

If $q \in M$, then $x^p \cdot x^{q+1} = x^p \cdot x^q \cdot x$ by **H2b**

$= x^{p+q} \cdot x$ because $q \in M$

$= x^{p+q+1}$ by **H2b.**

Therefore $q + 1 \in M$ and so, by induction, $x^p \cdot x^q = x^{p+q}$ whenever $q \geqslant 0$.

If $0 > q$, then $x^{p+q} \cdot (x^q)^- = x^{p+q} \cdot x^{-q}$ by **H5**

$= x^{p+q-q}$ because $-q \geqslant 0$

$= x^p$.

Therefore $x^p \cdot x^q = x^{p+q}$.

7. Theorem: $(x^p)^q = x^{p \cdot q}$.

Proof: Let M be the set of q for which this is true for all p. $(x^p)^0 = 1 = x^0 = x^{p \cdot 0}$. Therefore $0 \in M$.

If $q \in M$, then $(x^p)^{q+1} = (x^p)^q \cdot x^p$ by **H2b**

$= x^{p \cdot q} \cdot x^p$ because $q \in M$

$= x^{p \cdot (q+1)}$ by **H6.**

Therefore $q + 1 \in M$ and so, by induction, $(x^p)^q = x^{p \cdot q}$ whenever $q \geqslant 0$.

If $0 > q$, then $(x^p)^q = ((x^p)^-)^{-q}$ by **H2c**

$= (x^{-p})^{-q}$ by **H5**

$= x^{p \cdot q}$ because $-q \geqslant 0$.

8. Definition (of 0^p): If p is a non-zero positive integer, then $0^p = 0$. It follows that if p and q are non-zero positive integers, then $0^{p+1} = 0^p \cdot 0$; $0^p \cdot y^p = (0 \cdot y)^p$; $x^p \cdot 0^p = (x \cdot 0)^p$; $0^p \cdot 0^q = 0^{p+q}$; and $(0^p)^q = 0^{p \cdot q}$.
In other words, the previous results are true whenever they have a meaning.

9. Lemma (to **J12**): *If* $\boldsymbol{b}$ *is any positive non-zero rational number, there is an integer* p *such that* $\boldsymbol{b} \geqslant \boldsymbol{2}^{-p}$.

Proof: Let M be the set of integral rational numbers $\boldsymbol{q}$ for which there is a p such that $\boldsymbol{2}^p \geqslant \boldsymbol{q}$. $\boldsymbol{0} \in M$, because $\boldsymbol{2}^1 \geqslant \boldsymbol{0}$; and $\boldsymbol{1} \in M$, because $\boldsymbol{2}^1 \geqslant \boldsymbol{1}$.

If $\boldsymbol{q} \in M$ and $\boldsymbol{q} \geqslant \boldsymbol{1}$, let $\boldsymbol{2}^p \geqslant \boldsymbol{q}$. Then

$$\boldsymbol{2}^{p+1} = \boldsymbol{2} \cdot \boldsymbol{2}^p \geqslant \boldsymbol{2} \cdot \boldsymbol{q} = \boldsymbol{q} + \boldsymbol{q} \geqslant \boldsymbol{q} + \boldsymbol{1}$$

Therefore $\boldsymbol{q} + \boldsymbol{1} \in M$, and so, by induction, M contains all positive integral rational numbers.

Now, by **G8**, there is a positive integral rational $\boldsymbol{r}$ such that $\boldsymbol{r} \geqslant \boldsymbol{b}^{-1}$. By the above result, there is a p such that $\boldsymbol{2}^p \geqslant \boldsymbol{r}$. Then $\boldsymbol{b} \geqslant \boldsymbol{r}^{-1} \geqslant \boldsymbol{2}^{-p}$.

10. Theorem: *If* p *is a positive non-zero integer and* $x > y \geqslant 0$, *then* $x^p > y^p \geqslant 0$.

Proof: $x^1 > y^1 \geqslant 0$.

If $x^r > y^r \geqslant 0$, then $x^{r+1} = x^r \cdot x > y^r \cdot y = y^{r+1}$; and $y^{r+1} = y^r \cdot y \geqslant 0$. Therefore, by induction, $x^p > y^p \geqslant 0$ for every positive non-zero integer p.

11. Theorem: *If* $x \geqslant 0$, $y \geqslant 0$, $p \neq 0$, *and* $x^p = y^p$, *then* $x = y$.

Proof: (I). If $p > 0$, we cannot have $x > y$, because then $x^p > y^p$, by **H10**, and so $x^p \neq y^p$. Similarly, we cannot have $y > x$. Therefore $x = y$.

(II). If $0 > p$, then $(x^{-p})^{-} = x^p = y^p = (y^{-p})^{-}$.

Therefore $x^{-p} = y^{-p}$. Therefore, by (I), $x^{-} = y^{-}$, and so $x = y$.

12. Theorem: *If* $x > 1$ *and* $p > q$, *then* $x^p > x^q$.

Proof: $x^1 = x > 1$. If $x^r > 1$, then $x^{r+1} = x^r \cdot x > 1 \cdot 1 = 1$. Therefore, by induction, $x^r > 1$ for every positive non-zero integer r. In particular, $x^{p-q} > 1$. Therefore $x^p = x^{p-q} \cdot x^q > 1 \cdot x^q = x^q$.

13. Definition (of the phrase "*for all large . . .*"): If $\boldsymbol{S}(p)$ is a statement about an integer p, then $\boldsymbol{S}(p)$ is true *for all large* p if there is a n such that $\boldsymbol{S}(p)$ is true whenever $p \geqslant n$. Similarly $\boldsymbol{T}(p, q)$ is true *for all large* p *and* q if there is a n such that $\boldsymbol{T}(p, q)$ is true whenever $p \geqslant n$ and $q \geqslant n$.

Lemma (to many theorems in Chapter **I**): *If* $\boldsymbol{S}(p)$ *is true for all large* p *and if* $\boldsymbol{T}(p)$ *is true for all large* p, *then* $\boldsymbol{S}(p)$-*and*-$\boldsymbol{T}(p)$ *is true for all large* p. *If* $\boldsymbol{S}(p, q)$ *is true for all large* p *and* q *and if* $\boldsymbol{T}(p, q)$ *is true for all large* p *and* q, *then* $\boldsymbol{S}(p, q)$-*and*-$\boldsymbol{T}(p, q)$ *is true for all large* p *and* q.

Proof: Let $\boldsymbol{S}(p)$ be true whenever $p \geqslant m$ and $\boldsymbol{T}(p)$ be true whenever $p \geqslant n$. Let l be the greater of m and n. Then $\boldsymbol{S}(p)$ and $\boldsymbol{T}(p)$ are both true whenever $p \geqslant l$; that is, for all large p the statement $\boldsymbol{S}(p)$-and-$\boldsymbol{T}(p)$ is true. Similarly for $\boldsymbol{S}(p, q)$-and-$\boldsymbol{T}(p, q)$.

EXERCISES H

1. Prove that if $a \neq 0$ there is an element x such that $a \cdot x^2 + 2h \cdot x + b = 0$ if and only if there is an element y such that $y^2 = h^2 - a \cdot b$.
2. Prove that if p is an integer and $0 \geqslant x$, then $x^{2p} \geqslant 0 \geqslant x^{2p+1}$ (unless $x = p = 0$).
3. Prove that if $1 + h \geqslant 0$ and p is a positive integer, then $(1 + h)^p \geqslant 1 + p \cdot h$.
4. Prove that if $\boldsymbol{x}$ is a rational number other than $\boldsymbol{0}$, $\boldsymbol{1}$, or $-\boldsymbol{1}$, and $\boldsymbol{k}$ is any rational number, then there is an integer p such that $\boldsymbol{x}^p \geqslant \boldsymbol{k}$.

5*. t is an integer and $t \geqslant 2$. Prove that if a is an integer and $a > 0$, then there are integers $x_1 \dots x_n$ for which $t > x_p \geqslant 0$, $x_n > 0$, and $a = \sum_{p=0}^{n} x_p \cdot t^p$; and that if also $a = \sum_{p=0}^{m} y_p \cdot t^p$, where the y_p are integers, $t > y_p \geqslant 0$, and $y_m > 0$; then $m = n$, and $x_p = y_p$ for each p.

6. Prove that it is impossible to define 0^p for all p in such a way that **H2a** and **b** are true for all x.

CHAPTER I

Cauchy Numbers

1. Notation: A succession (see **A1**) has a member which is not a successor, by **A1b**: we shall write it with a suffix 1, its successor with a suffix 2, and so on. We shall write the succession whose members are $x_1, x_2, \ldots$ as $\{x\}$. If we have to consider a second succession we shall use another letter; e.g. we may denote it by $\{y\}$ and its elements by $y_1, y_2, \ldots$ We shall use l, m, n, p, q, r, s, and t to denote integers.

2. Definition (of *Cauchy sequence*): A succession $\{x\}$ of elements of an ordered field is a *Cauchy sequence* if, for each positive non-zero b, $b \geqslant | x_p - x_q |$ for all large p and q (see **H13**).

3. Definition (of *Cauchy number*): If $\{x\}$ is a Cauchy sequence, the Cauchy number $\mathbf{x}$ is the set of all Cauchy sequences $\{y\}$ with the following property:

For each positive non-zero b, $b \geqslant | x_p - y_p |$ for all large p.

Note: by **F23**, this implies that, for each positive non-zero b, $\frac{1}{2}b \geqslant | x_p - y_p |$ for all large p.

4. Theorem: *The Cauchy sequence* $\{x\}$ *is in the Cauchy number* $\mathbf{x}$ *and in no other Cauchy number.*

Proof: If b is positive and non-zero, then

$$b \geqslant 0 = | x_p - x_p | \text{ for every } p.$$

Therefore $\qquad \{x\} \in \mathbf{x}$ by definition **I3**.

If $\{x\} \in \mathbf{y}$, let $\{u\}$ be any element of $\mathbf{x}$ and $\{v\}$ be any element of $\mathbf{y}$.

By note **I3**,

(i) $$\tfrac{1}{2}b \geqslant | y_p - x_p |$$

for all large p;

(ii) $$\tfrac{1}{2}b \geqslant | x_p - u_p |$$

for all large p; and

(iii) $$\tfrac{1}{2}b \geqslant | y_p - v_p |$$

for all large p.

By **H13**, applied twice, (i), (ii), and (iii) are all true for all large p. Then, for all large p,

$$\begin{aligned} b &\geqslant | x_p - u_p | + | y_p - x_p | && \text{by (i) and (ii)} \\ &\geqslant | y_p - u_p | && \text{by } \mathbf{F20}. \end{aligned}$$

Therefore $\{u\} \in \mathbf{y}$, by definition **I3**.

Also
$$\begin{aligned} b &\geqslant | y_p - x_p | + | y_p - v_p | && \text{by (i) and (iii)} \\ &= | x_p - y_p | + | y_p - v_p | && \text{by } \mathbf{F16} \\ &\geqslant | x_p - v_p | && \text{by } \mathbf{F20}. \end{aligned}$$

Therefore $\{v\} \in \mathbf{x}$, by definition **I3**.

We have proved that every element of $\mathbf{x}$ is an element of $\mathbf{y}$ and every element of $\mathbf{y}$ is an element of $\mathbf{x}$. Therefore $\mathbf{x} = \mathbf{y}$, and so $\mathbf{x}$ is the only Cauchy number containing $\{x\}$.

5. Lemma (to definition **I6**): *If* $\{u\} \in \mathbf{x}$ *and* $\{v\} \in \mathbf{y}$, *then* $\{u + v\}$ *and* $\{x + y\}$ *are Cauchy sequences and are both in the same Cauchy number.*

Note: $\{x + y\}$ denotes, of course, the succession whose p^{th} member is $x_p + y_p$.

Proof: If $b > 0$, then, by definition **I2** (and lemma **H13**),

$$\tfrac{1}{2}b \geqslant | x_p - x_q | \text{ and } \tfrac{1}{2}b \geqslant | y_p - y_q |$$

for all large p and q. Then

$$\begin{aligned} b &\geqslant | x_p - x_q + y_p - y_q | \qquad \text{by } \mathbf{F20} \\ &= | (x_p + y_p) - (x_q + y_q) |. \end{aligned}$$

Therefore $\{x + y\}$ is a Cauchy sequence, by definition **I2**. Similarly $\{u + v\}$ is a Cauchy sequence.

If $\{u\} \in \mathbf{x}$ and $\{v\} \in \mathbf{y}$, then, by definition **I3**,

$$\tfrac{1}{2}b \geqslant | x_p - u_p | \text{ and } \tfrac{1}{2}b \geqslant | y_p - v_p |$$

for all large p. Then

$$b \geqslant | (x_p + y_p) - (u_p + v_p) |.$$

Therefore, by definition **I3**, $\{u + v\}$ is in the Cauchy number which, by **I4**, contains $\{x + y\}$.

6. Definition (of *sum* of Cauchy numbers): $\mathbf{x} + \mathbf{y}$ is the Cauchy number which contains $\{u + v\}$ whenever $\{u\} \in \mathbf{x}$ and $\{v\} \in \mathbf{y}$.

Note: That there is such a Cauchy number was proved in **I5.**

7. Lemma (to **I9**): $\{x + (y + z)\} \in \mathbf{x} + (\mathbf{y} + \mathbf{z})$

and $\{(x + y) + z\} \in (\mathbf{x} + \mathbf{y}) + \mathbf{z}$.

Proof: By definition **I6**, $\{x + (y + z)\} \in \mathbf{x} + \mathbf{w}$ where $\mathbf{w}$ is the Cauchy number containing $\{y + z\}$. Then $\mathbf{w} = \mathbf{y} + \mathbf{z}$, by definition **I6** again.

Similarly for the other.

8. Theorem: $\mathbf{x} + \mathbf{y} = \mathbf{y} + \mathbf{x}$.

Proof: Let $\{x\} \in \mathbf{x}$ and $\{y\} \in \mathbf{y}$.

Then $\{y + x\} = \{x + y\}$

$\in \mathbf{x} + \mathbf{y}$ by definition **I6.**

But $\{y + x\} \in \mathbf{y} + \mathbf{x}$ by definition **I6.**

Therefore $\mathbf{x} + \mathbf{y} = \mathbf{y} + \mathbf{x}$ by **I4.**

9. Theorem: $(\mathbf{x} + \mathbf{y}) + \mathbf{z} = \mathbf{x} + (\mathbf{y} + \mathbf{z})$.

Proof: Let $\{x\} \in \mathbf{x}$, $\{y\} \in \mathbf{y}$, and $\{z\} \in \mathbf{z}$.

Then $\{(x + y) + z\} = \{x + (y + z)\}$

$\in \mathbf{x} + (\mathbf{y} + \mathbf{z})$ by **I7.**

But $\{(x + y) + z\} \in (\mathbf{x} + \mathbf{y}) + \mathbf{z}$ by **I7.**

Therefore $(\mathbf{x} + \mathbf{y}) + \mathbf{z} = \mathbf{x} + (\mathbf{y} + \mathbf{z})$ by **I4.**

10. Theorem: *If $x_p = a$ for every p, then $\{x\}$ is a Cauchy sequence.*

Proof: If b is positive and non-zero, then

$$b \geqslant 0 = | x_p - x_q | \text{ whenever } p \geqslant 1 \text{ and } q \geqslant 1.$$

Therefore $\{x\}$ is a Cauchy sequence, by definition **I2.**

Notation: We denote this Cauchy sequence by $[a]$.

11. Definition (of **0**): **0** is the Cauchy number containing [0] (which is a Cauchy sequence by **I10**).

12. Theorem: $\mathbf{0} + \mathbf{x} = \mathbf{x}$.

Proof: Let $\{x\} \in \mathbf{x}$. Then $\{x\} = \{0 + x\} \in \mathbf{0} + \mathbf{x}$, by definition **I6**. Therefore $\mathbf{x} = \mathbf{0} + \mathbf{x}$, by **I4**.

13. Theorem: *If* $\{x\}$ *is a Cauchy sequence, so is* $\{-x\}$.

Proof: If b is positive and non-zero, then, by definition **I2**,

$$b \geqslant | x_q - x_p | \quad \text{for all large } p \text{ and } q$$
$$= | -x_p - (-x_q) | .$$

Therefore $\{-x\}$ is a Cauchy sequence, by definition **I2**.

14. Theorem: *If* $\mathbf{a}$ *is any Cauchy number there is a* $\mathbf{x}$ *such that* $\mathbf{a} + \mathbf{x} = \mathbf{0}$.

Proof: Let $\{a\} \in \mathbf{a}$ and $\mathbf{x}$ be the Cauchy number containing $\{-a\}$, which is a Cauchy sequence by **I13**. Then

$$[0] = \{a + (-a)\} \in \mathbf{a} + \mathbf{x} \qquad \text{by definition } \mathbf{I6}.$$

Therefore $\mathbf{0} = \mathbf{a} + \mathbf{x}$ by **I4**.

15. Lemma (to **I16** and **J2**): *If* $\{x\}$ *is a Cauchy sequence, there is a positive non-zero* k *such that* $k \geqslant | x_p |$ *for all large* p.

Proof: By definitions **I2** and **H13** there is a n such that

$$1 \geqslant | x_p - x_q | \qquad \text{whenever } p \geqslant n \text{ and } q \geqslant n.$$

Therefore $1 \geqslant | x_p - x_n | \qquad$ whenever $p \geqslant n$.

Then

$$1 + | x_n | \geqslant | x_p - x_n | + | x_n |$$
$$\geqslant | x_p | \qquad \text{by } \mathbf{F20}.$$

Then $1 + | x_n |$ is the desired k.

16. Lemma (to definition **I17**): *If* $\{u\} \in \mathbf{x}$ *and* $\{v\} \in \mathbf{y}$, *then* $\{u \cdot v\}$ *and* $\{x \cdot y\}$ *are Cauchy sequences and are both in the same Cauchy number.*

Proof: By **I15** there is a positive non-zero k such that

(i) $$k \geqslant | x_q |$$

for all large q, and there is a positive non-zero f such that

(ii) $$f \geqslant | y_p |$$

for all large p.

If b is positive and non-zero then, by definition **I2** and lemma **H13**,

(iii) $$\tfrac{1}{2}b\cdot f^{-1} \geqslant | x_p - x_q |$$

and

(iv) $$\tfrac{1}{2}b\cdot k^{-1} \geqslant | y_p - y_q |$$

for all large p and q. Then, for all large p and q,

$$\tfrac{1}{2}b \geqslant | y_p | \cdot | x_p - x_q | \qquad \text{by (ii) and (iii)}$$
$$= | x_p \cdot y_p - x_q \cdot y_p |$$

and $$\tfrac{1}{2}b \geqslant | x_q | \cdot | y_p - y_q | \qquad \text{by (i) and (iv)}$$
$$= | x_q \cdot y_p - x_q \cdot y_q | .$$

Therefore $$b \geqslant | x_p \cdot y_p - x_q \cdot y_q | \qquad \text{by } \mathbf{F20}.$$

Therefore $\{x\cdot y\}$ is a Cauchy sequence, by definition **I2**.
Similarly, so is $\{u\cdot v\}$.

By **I15** there is a positive non-zero h such that

(v) $$h \geqslant | u_p |$$

for all large p.
By definition **I3** and because $\{u\} \in \mathbf{x}$ and $\{v\} \in \mathbf{y}$,

(vi) $$\tfrac{1}{2}b\cdot f^{-1} \geqslant | x_p - u_p |$$

and

(vii) $$\tfrac{1}{2}b\cdot h^{-1} \geqslant | y_p - v_p | \qquad \text{for all large } p.$$

Then $$\tfrac{1}{2}b \geqslant | y_p | \cdot | x_p - u_p | \qquad \text{by (ii) and (vi)}$$
$$= | x_p \cdot y_p - y_p \cdot u_p |$$

and $$\tfrac{1}{2}b \geqslant | u_p | \cdot | y_p - v_p | \qquad \text{by (v) and (vii)}$$
$$= | y_p \cdot u_p - u_p \cdot v_p | .$$

Therefore $$b \geqslant | x_p \cdot y_p - u_p \cdot v_p | .$$

Therefore, by definition **I3**, $\{u\cdot v\}$ is in the Cauchy number which, by **I4**, contains $\{x\cdot y\}$.

17. Definition (of *product* of Cauchy numbers): $\mathbf{x}\cdot\mathbf{y}$ is the Cauchy number which contains $\{u\cdot v\}$ whenever $\{u\} \in \mathbf{x}$ and $\{v\} \in \mathbf{y}$.

Note: That there is such a Cauchy number was proved in **I16**.

18. Lemma (to **I20**): $\{x\cdot(y\cdot z)\} \in \mathbf{x}\cdot(\mathbf{y}\cdot\mathbf{z})$ and $\{(x\cdot y)\cdot z\} \in (\mathbf{x}\cdot\mathbf{y})\cdot\mathbf{z}$.

Proof: By definition **I17**, $\{x\cdot(y\cdot z)\} \in \mathbf{x}\cdot\mathbf{w}$ where $\mathbf{w}$ is the Cauchy number containing $\{y\cdot z\}$. Then $\mathbf{w} = \mathbf{y}\cdot\mathbf{z}$ by definition **I17** again. Similarly for the other.

19. Theorem: $\mathbf{x} \cdot \mathbf{y} = \mathbf{y} \cdot \mathbf{x}$.

Proof: Let $\{x\} \in \mathbf{x}$ and $\{y\} \in \mathbf{y}$.

Then	$\{y \cdot x\} = \{x \cdot y\}$	
	$\in \mathbf{x} \cdot \mathbf{y}$	by definition **I17**.
But	$\{y \cdot x\} \in \mathbf{y} \cdot \mathbf{x}$	by definition **I17**.
Therefore	$\mathbf{x} \cdot \mathbf{y} = \mathbf{y} \cdot \mathbf{x}$	by **I4**.

20. Theorem: $(\mathbf{x} \cdot \mathbf{y}) \cdot \mathbf{z} = \mathbf{x} \cdot (\mathbf{y} \cdot \mathbf{z})$.

Proof: Let $\{x\} \in \mathbf{x}$, $\{y\} \in \mathbf{y}$, and $\{z\} \in \mathbf{z}$.

Then	$\{(x \cdot y) \cdot z\} = \{x \cdot (y \cdot z)\}$	
	$\in \mathbf{x} \cdot (\mathbf{y} \cdot \mathbf{z})$	by **I18**.
But	$\{(x \cdot y) \cdot z\} \in (\mathbf{x} \cdot \mathbf{y}) \cdot \mathbf{z}$	by **I18**.
Therefore	$(\mathbf{x} \cdot \mathbf{y}) \cdot \mathbf{z} = \mathbf{x} \cdot (\mathbf{y} \cdot \mathbf{z})$	by **I4**.

21. Lemma (to **I24**, **31**, and **33**): *If* $\{x\}$ *is a Cauchy sequence and* $\mathbf{x} \neq \mathbf{0}$, *then either* (i) *there is a* k *such that* $x_p \geqslant k > 0$ *for all large* p *or* (ii) *there is a* k *such that* $-x_p \geqslant k > 0$ *for all large* p.

Proof: If $b > 0$, there is, by definitions **I2** and **H13**, a n such that

(iii) $\quad \frac{1}{2}b \geqslant | x_p - x_q |$ whenever $p \geqslant n$ and $q \geqslant n$.

If (i) is false, there is a r such that

(iv) $\quad \frac{1}{2}b \geqslant x_r \quad$ and $\quad r \geqslant n$.

If (ii) is false, there is a s such that

(v) $\quad \frac{1}{2}b \geqslant -x_s \quad$ and $\quad s \geqslant n$.

Therefore, if (i) and (ii) are false and $p \geqslant n$,

	$\frac{1}{2}b \geqslant \mid x_p - x_r \mid$	by (iii) and because $r \geqslant n$.
Then	$b \geqslant \mid x_p - x_r \mid + x_r$	by (iv)
	$\geqslant x_p - x_r + x_r$	
	$= x_p$.	
Similarly,	$b \geqslant \mid x_s - x_p \mid - x_s$	by (iii) and (v)
	$\geqslant x_s - x_p - x_s$	
	$= -x_p$.	

But $| x_p |$ is either x_p or $-x_p$. Therefore

$$b \geqslant | x_p | = | x_p - 0 | \quad \text{whenever} \quad p \geqslant n.$$

Therefore $\mathbf{x} = \mathbf{0}$, by definitions **I11** and **I3**. Therefore if $\mathbf{x} \neq \mathbf{0}$, either (i) or (ii) must be true.

22. Definition (of **1**): **1** is the Cauchy number containing [1], (which is a Cauchy sequence by **I10**). Clearly $\mathbf{1} \neq \mathbf{0}$.

23. Theorem: $\mathbf{1 \cdot x} = \mathbf{x}$.

Proof: Let $\{x\} \in \mathbf{x}$. Then

$$\{x\} = \{1 \cdot x\}$$

$$\in \mathbf{1 \cdot x} \quad \text{by definitions } \mathbf{I17} \text{ and } \mathbf{I22}.$$

Therefore $\mathbf{x} = \mathbf{1 \cdot x}$ by **I4**.

24. Lemma (to **I25**): *If* $\{x\}$ *is a Cauchy sequence and* $\mathbf{x} \neq \mathbf{0}$, *there is a Cauchy sequence* $\{\hat{x}\}$ *such that*

(i) $$\hat{\mathbf{x}} = \mathbf{x}$$

and

(ii) $$\hat{x}_p \neq 0, \textit{ for every } p.$$

Proof: By **I21**, there are a n and a k such that $|x_p| \geqslant k > 0$ whenever $p \geqslant n$. Let $\hat{x}_p$ be x_p when $p \geqslant n$ and k otherwise.

(i) If $p \geqslant n$ and b is positive and non-zero, then

$$b \geqslant 0 = |\hat{x}_p - x_p|.$$

Therefore $\mathbf{x} = \hat{\mathbf{x}}$ by **I4** and definition **I3**.

(ii) If $p \geqslant n$, then $|x_p| \geqslant k$. But $k > 0$. Therefore $\hat{x}_p = x_p \neq 0$.

25. Lemma (to **I26**): *If* $\{x\}$ *is a Cauchy sequence and* $\mathbf{x} \neq \mathbf{0}$, *then* $\{\hat{x}^{-1}\}$ *is a Cauchy sequence.*

Note: $\{\hat{x}\}$ is defined in **I24**.

Proof: By **I21**, there is a k such that

(i) $$|\hat{x}_p| \geqslant k > 0$$

for all large p. But $\hat{x}_p \neq 0$, by **I24**. Therefore

(ii) $$k^{-1} \geqslant |\hat{x}_p^{-1}|$$

for all large p, by (i), **F14**, and **F18**.

If $b > 0$, then, by definition **I2**,

(iii) $$b \cdot k^2 \geqslant |\hat{x}_p - \hat{x}_q|$$

for all large p and q. Then

$$b \geqslant |\hat{x}_q^{-1}| \cdot |\hat{x}_p - \hat{x}_q| \cdot |\hat{x}_p^{-1}| \quad \text{by (ii) and (iii)}$$
$$= |\hat{x}_q^{-1} - \hat{x}_p^{-1}| \quad \text{by } \mathbf{F17}.$$

Therefore, by definition **I2**, $\{\hat{x}^{-1}\}$ is a Cauchy sequence.

26. Theorem: *If* $\mathbf{a} \neq \mathbf{0}$, *there is a* $\mathbf{x}$ *for which* $\mathbf{a}\cdot\mathbf{x} = \mathbf{1}$.

Proof: Let $\{a\} \in \mathbf{a}$, $\{\hat{a}\}$ be as in **I24**, and $\mathbf{x}$ be the Cauchy number containing $\{\hat{a}^{-1}\}$, which is a Cauchy sequence by **I25**.

Then	$[1] = \{\hat{a}\cdot\hat{a}^{-1}\}$	
	$\in \hat{\mathbf{a}}\cdot\mathbf{x}$	by definition **I17**
	$= \mathbf{a}\cdot\mathbf{x}$	by **I24**.
But	$[1] \in \mathbf{1}$	by definition **I22**.
Therefore	$\mathbf{a}\cdot\mathbf{x} = \mathbf{1}$	by **I4**.

27. Theorem: $\mathbf{x}\cdot(\mathbf{y} + \mathbf{z}) = \mathbf{x}\cdot\mathbf{y} + \mathbf{x}\cdot\mathbf{z}$.

Proof: Let $\{x\} \in \mathbf{x}$, $\{y\} \in \mathbf{y}$, and $\{z\} \in \mathbf{z}$.

Then	$\{x\cdot(y + z)\} = \{x\cdot y + x\cdot z\}$	
	$\in \mathbf{x}\cdot\mathbf{y} + \mathbf{x}\cdot\mathbf{z}$	by definitions **I6** and **I17**.
But	$\{x\cdot(y + z)\} \in \mathbf{x}\cdot(\mathbf{y} + \mathbf{z})$	by definitions **I17** and **I6**.
Therefore	$\mathbf{x}\cdot(\mathbf{y} + \mathbf{z}) = \mathbf{x}\cdot\mathbf{y} + \mathbf{x}\cdot\mathbf{z}$	by **I4**.

28. Theorem: *The Cauchy numbers form a field.*

Proof: **D1a-k** follow from **I9, 8, 12, 14, 20, 19, 23, 26,** and **27** respectively.

$-\mathbf{x}$ is the Cauchy number containing $\{-x\}$ where $\{x\} \in \mathbf{x}$, by **I14**.
$\mathbf{x}^{-1}$ is the Cauchy number containing $\{\hat{x}^{-1}\}$ where $\{x\} \in \mathbf{x}$, by **I26**.
i is **1** and o is **0**.

Definition (of *Cauchy field*): This field is the *Cauchy field* of the ordered field from which the elements of the Cauchy sequences are taken.

29. Definition (of *positive* Cauchy number): A Cauchy number is *positive* if it contains a Cauchy sequence $\{x\}$ such that $x_p \geqslant 0$ for all large p.

30. Theorem: **0** *is positive.*

Proof: **0** contains [0], and $0 \geqslant 0$.

31. Lemma (to **I32**, **38**, and **39**): *If* $\mathbf{x}$ *is positive and non-zero and* $\{x\} \in \mathbf{x}$, *then there is a* k *such that* $x_p \geqslant k > 0$ *for all large* p.

Proof: If $\mathbf{x}$ is positive then there is a $\{y\}$ in $\mathbf{x}$ such that

(i) $$y_p \geqslant 0$$

for all large p, by definition **I29**.

If there is a k such that

(ii) $$-x_p \geqslant k > 0$$

for all large p; then, by definition **I3**,

$$\begin{aligned} \tfrac{1}{2}k &\geqslant |y_p - x_p| \\ &\geqslant y_p - x_p \\ &\geqslant k \qquad \text{by (i) and (ii).} \end{aligned}$$

Therefore $0 \geqslant \frac{1}{2}k$, which is impossible, by (ii). Therefore there can be no k for which $-x_p \geqslant k > 0$ for all large p; and so, by **I21**, there is a k such that $x_p \geqslant k > 0$ for all large p.

32. Theorem: *If* $\mathbf{x}$ *and* $-\mathbf{x}$ *are both positive, then* $\mathbf{x} = \mathbf{0}$.

Proof: If $\mathbf{x}$ and $-\mathbf{x}$ are positive and non-zero, let $\{x\} \in \mathbf{x}$. Then $\{-x\} \in -\mathbf{x}$. Therefore, by **I31**, there is a k such that $x_p \geqslant k > 0$ for all large p, and a h such that $-x_p \geqslant h > 0$ for all large p. Then $0 = -x_p + x_p \geqslant k + h$, which is impossible because $k > 0$ and $h > 0$. Therefore $\mathbf{x}$ is *not* non-zero.

33. Theorem: *If* $\mathbf{x}$ *is any Cauchy number, then either* $\mathbf{x}$ *or* $-\mathbf{x}$ *is positive.*

Proof: If $\mathbf{x}$ is not positive, then $\mathbf{x} \neq \mathbf{0}$, by **I30**. Let $\{x\} \in \mathbf{x}$. Then, by **I21**, either (i) $x_p \geqslant 0$ for all large p or (ii) $-x_p \geqslant 0$ for all large p. (i) is false if $\mathbf{x}$ is not positive, by definition **I29**. Therefore (ii) is true, and so $-\mathbf{x}$ is positive.

34. Theorem: *If* $\mathbf{x}$ *and* $\mathbf{y}$ *are positive, so are* $\mathbf{x} + \mathbf{y}$ *and* $\mathbf{x}\cdot\mathbf{y}$.

Proof: Let $\{x\} \in \mathbf{x}$, $\{y\} \in \mathbf{y}$, and $\{x\}$ and $\{y\}$ have property **I29**. Then $\{x + y\}$ and $\{x \cdot y\}$ have property **I29** by **F1c**. Therefore $\mathbf{x} + \mathbf{y}$ and $\mathbf{x}\cdot\mathbf{y}$ are positive, by definition **I29**.

35. Theorem: *The Cauchy numbers form an ordered field.*

Proof: Let P be the set of positive Cauchy numbers. Then **F1a**

b, and **c** follow from **I33**, **32**, and **34**. We may now apply all the theorems and definitions of **F** to the Cauchy numbers. In particular, we may define $|\mathbf{x}|$ and the relation $\mathbf{x} \geqslant \mathbf{y}$. We can then define Cauchy sequences of Cauchy numbers.

36. Notation: $\tilde{x}$ is the Cauchy number containing $[x]$, (which was defined in **I10**).

37. Theorem: *The correspondence*

$$\tilde{x} \longleftrightarrow x$$

is an isomorphism between Cauchy numbers of the form $\tilde{x}$ *and the elements of the original field.*

Proof: If $\tilde{x} = \tilde{y}$, then $[x]$ and $[y]$ are in the same Cauchy number, by **I36**. If $x \neq y$, let $|x - y| = b$. Then b is positive and non-zero. Therefore so is $\frac{1}{2}b$.

Therefore $\quad \frac{1}{2}b \geqslant |x - y| \quad$ by definition **I3**

$\qquad\qquad\quad = b.$

Therefore $\quad 0 \geqslant \frac{1}{2}b \quad$ which is impossible.

Therefore x cannot be different from y.

Therefore the correspondence is a matching.

$[x + y] \in (x + y)^{\sim} \quad$ by **I36**.

$[x + y] \in \tilde{x} + \tilde{y} \quad$ by definition **I6**.

Therefore $\quad (x + y)^{\sim} = \tilde{x} + \tilde{y}$.

Similarly, by **I36** and definition **I17**,

$$(x{\cdot}y)^{\sim} = \tilde{x}{\cdot}\tilde{y}.$$

Therefore the matching is an isomorphism.

Note: It follows, of course, that $-\tilde{x} = (-x)^{\sim}$; $(\tilde{x}^{-1}) = (x^{-1})^{\sim}$; and so on.

38. Theorem: $\tilde{x}$ *is positive if and only if* x *is positive.*

Proof: If $\tilde{x}$ is zero, then, because $[x] \in \tilde{x}$, $x = 0$. If $\tilde{x}$ is positive and non-zero, then $x \geqslant 0$ by **I31**.

On the other hand, if x is positive then $[x]$ has property **I29** and so $\tilde{x}$ is positive.

Note: It follows of course that $\tilde{x} \geqslant \tilde{y}$ if and only if $x \geqslant y$; that $|\tilde{x}| = |x|^{\sim}$; and so on.

Definition (of *order-isomorphism*): An isomorphism $x \longleftrightarrow y$ between two ordered fields is an *order-isomorphism* if x is positive whenever y is positive and y is positive whenever x is positive.

39. Lemma (to **I40** and **41**): *If* $\mathbf{x} > \mathbf{y}$, *there is a* z *such that* $\mathbf{x} \geqslant \tilde{z} \geqslant \mathbf{y}$.

Proof: Let $\{x\} \in \mathbf{x}$ and $\{y\} \in \mathbf{y}$. Then $\{x - y\} \in \mathbf{x} - \mathbf{y}$, which is positive and non-zero; and so, by **I31**, there is a positive non-zero b such that

(i) $$x_p - y_p \geqslant 2b$$

for all large p. By definition **I2**, $b \geqslant | y_p - y_q |$ for all large p and q. Then

(ii) $$y_p - y_q \geqslant -b.$$

Let $z = y_l + b$, where l is such that (i) and (ii) are true whenever $p \geqslant l$ and $q \geqslant l$. Then, whenever $p \geqslant l$,

$$\begin{aligned} x_p - z &= (x_p - y_p) + (y_p - y_l) - b \\ &\geqslant 2b - b - b \qquad \text{by (i) and (ii)} \\ &= 0. \end{aligned}$$

Therefore $\mathbf{x} - \tilde{z}$ is positive and so $\mathbf{x} \geqslant \tilde{z}$.

Also
$$\begin{aligned} z - y_p &= y_l - y_p + b \\ &\geqslant -b + b \qquad \text{by (ii)} \\ &= 0. \end{aligned}$$

Therefore $\tilde{z} - \mathbf{y}$ is positive, and so $\tilde{z} \geqslant \mathbf{y}$.

40. Lemma (to **I41** and **42**): *If* $\{x\} \in \mathbf{x}$, *and if* $\mathbf{b} > \mathbf{0}$, *then* $\mathbf{b} \geqslant |\tilde{x}_p - \mathbf{x}|$ *for all large* p.

Proof: By **I39**, there is a g such that $\mathbf{b} \geqslant \tilde{g} \geqslant \frac{1}{2}\mathbf{b}$. By definition **I2**,

(i) $$g \geqslant | x_p - x_q | \quad \text{for all large } p \text{ and } q.$$

Now $\tilde{g} - (\tilde{x}_p - \mathbf{x})$ contains the succession whose q^{th} member is $g - (x_p - x_q)$, and, by (i), $g - (x_p - x_q) \geqslant 0$ for all large p and q. Therefore, by definition **I29**, $\tilde{g} - (\tilde{x}_p - \mathbf{x}) \geqslant \mathbf{0}$ for all large p.

Similarly $$\tilde{g} - (\mathbf{x} - \tilde{x}_p) \geqslant \mathbf{0} \text{ for all large } p.$$

Therefore $\tilde{g} \geqslant |\tilde{x}_p - \mathbf{x}|$ for all large p. But $\mathbf{b} \geqslant \tilde{g}$.

41. Lemma (to **I42**): *If* $\{\mathbf{x}\}$ *is a Cauchy sequence of Cauchy numbers, there is a Cauchy sequence* $\{\tilde{z}\}$ *with the following property*:

$$\text{If } \mathbf{b} > \mathbf{0}, \text{ then } \mathbf{b} \geqslant |\, \mathbf{x}_p - \tilde{z}_p \,| \quad \text{for all large } p.$$

Proof: If there is a Cauchy number $\mathbf{z}$ such that $\mathbf{x}_p = \mathbf{z}$ for all large p, let $\{z\} \in \mathbf{z}$. Then if $\mathbf{b} > \mathbf{0}$

$$\begin{aligned} \mathbf{b} &\geqslant |\, \mathbf{z} - \tilde{z}_p \,| \quad \text{for all large } p, \text{ by } \mathbf{I40} \\ &= |\, \mathbf{x}_p - \tilde{z}_p \,|. \end{aligned}$$

If there is no such $\mathbf{z}$, we can define a succession $\{\mathbf{y}\}$ as follows: (i) $\mathbf{y}_1$ is $\mathbf{x}_1$; (ii) if $\mathbf{y}_m$ is defined as $\mathbf{x}_{p_m}$, then $\mathbf{y}_{m+1}$ is the first $\mathbf{x}_r$ after $\mathbf{x}_{p_m}$ for which $\mathbf{x}_r \neq \mathbf{y}_m$. Clearly $\{\mathbf{y}\}$ is a Cauchy sequence if $\{\mathbf{x}\}$ is. Therefore if $\mathbf{b}$ is positive and non-zero

$$\mathbf{b} \geqslant |\, \mathbf{y}_p - \mathbf{y}_{p+1} \,| \quad \text{for all large } p.$$

By **I39** there is, for each p, a z_p such that

$$\mathbf{x}_p + |\, \mathbf{y}_p - \mathbf{y}_{p+1} \,| \geqslant \tilde{z}_p \geqslant \mathbf{x}_p.$$

Then

$$\mathbf{b} \geqslant |\, \mathbf{y}_p - \mathbf{y}_{p+1} \,| \geqslant |\, \mathbf{x}_p - \tilde{z}_p \,| \quad \text{for all large } p.$$

42. We have extended an ordered field by the use of Cauchy sequences. It is natural to wonder what happens when we try to extend the Cauchy field by the same process. Let us call the numbers we get in this way super-Cauchy-numbers. The answer is that we get nothing essentially new: the field of super-Cauchy-numbers is order-isomorphic to the field of Cauchy numbers.

Proof: By **I37** and **38** the matching

$$\tilde{\mathbf{x}} \longleftrightarrow \mathbf{x}$$

is an order-isomorphism between super-Cauchy-numbers of the form $\tilde{\mathbf{x}}$ and the Cauchy numbers.

Let X be any super-Cauchy-number. Let $\{\mathbf{x}\} \in X$. Let $\{\tilde{z}\}$ be as in **I41**. Then $\{\tilde{z}\} \in X$, by definition **I3**. Let $\mathbf{z}$ be the Cauchy number which contains $\{z\}$. Then, by **I40** and definition **I3**, $[\mathbf{z}] \in X$. Therefore $\tilde{\mathbf{z}} = X$. Therefore every super-Cauchy-number is of the form $\tilde{\mathbf{x}}$, and so the matching is between the field of all super-Cauchy-numbers and the field of Cauchy numbers.

EXERCISES I

1. Prove, straight from definitions **I2** and **3**, that if $\mathbf{u} = \mathbf{v}$ and $\{x\}$ is a Cauchy sequence, then $\{u \cdot x\}$ and $\{v \cdot x\}$ are Cauchy sequences in the same Cauchy number. Hence give an alternative proof of **I16**.

2. A succession $\{x\}$ of elements of an ordered field has the property (α) that if b is positive and non-zero, then $b \geqslant | x_p |$ for all large p. Prove that $\{x\}$ is a Cauchy sequence, and that if $\{y\}$ is a Cauchy sequence then $\{x \cdot y\}$ has the property (α). Prove that $\{x + y\}$ has the property (α) if and only if $\{y\}$ has the property (α).

3. Prove that if $\{n_p\}$ is a succession of *distinct* whole numbers (that is, $n_p \neq n_q$ if $p \neq q$), if $b_p = a_{n_p}$ for every p, and if $\{a\}$ is a Cauchy sequence, then $\{b\}$ is a Cauchy sequence in the same Cauchy number.

4. Prove that if an ordered field F has the property that if x is any element of the field then there is an integer n (see **F21**) such that $n \geqslant x$, then so has its Cauchy field.

CHAPTER J

Real Numbers

1. Definition (of *real number*): A *real number* is an element of the Cauchy field (**I28**) of the field of rational numbers. It follows that the real numbers form an ordered field and that it contains a field order-isomorphic to the field of rational numbers, by **I35**, **37**, and **38**. The elements of this field (that is, elements of the form $\tilde{x}$, where x is a rational number) will be called *rational real numbers*. If x is an integral rational number, then $\tilde{x}$ will be called an *integral real number*. When there is no particular reason to emphasize the existence of the isomorphism (i.e. the fact that $\tilde{x}$ only *corresponds to* and is not *identical with* x), we shall abbreviate these terms to *rational number* and *integer* respectively.

Notation: We shall use l, m, n, p, q, r, s, and t for integers; other Roman letters for rational numbers, and bold-face letters for real numbers which are not necessarily rational; i.e. just as we abbreviate *rational real number* to *rational number*, so we abbreviate $\tilde{x}$ to x (and $\mathbf{0}$ to 0).

2. Theorem: *If* $\mathbf{x}$ *is any real number, there is an integral real number* $\tilde{n}$ *such that* $\tilde{n} \geqslant \mathbf{x}$.

Proof: In Chapter **I**, take the field to be the field of rational numbers. Then in **I15** the k is rational, because every x_p is.

Therefore for any Cauchy sequence $\{x\}$ of rational numbers there is a positive non-zero rational number k such that $k - x_p \geqslant 0$ for all large p. Let $\tilde{k}$ be the real number containing $[k]$ and $\mathbf{x}$ the real number containing $\{x\}$. Then $\tilde{k} - \mathbf{x}$ contains the succession whose p^{th} element is $k - x_p$, and so is positive, by definition **I29**; that is, $\tilde{k} \geqslant \mathbf{x}$.

By **G8**, there is an integral rational n such that $n \geqslant k$. Then $\tilde{n} \geqslant \tilde{k} \geqslant \mathbf{x}$.

3. Definition (of *limit*): If, for each positive non-zero $\mathbf{b}$,

$$\mathbf{b} \geqslant |\, \mathbf{x}_p - \mathbf{a} \,| \qquad \text{for all large } p,$$

then $\mathbf{a}$ is the *limit* of $\{\mathbf{x}\}$. This is written as

$$\lim\{\mathbf{x}\} = \mathbf{a}.$$

Note: Clearly $\lim[\mathbf{y}] = \mathbf{y}$. And clearly a succession cannot have more than one limit.

4. Theorem: *A succession of real numbers has a limit if and only if it is a Cauchy sequence.*

Proof: Let $\{\mathbf{x}\}$ be a Cauchy sequence and let $\{\tilde{z}\}$ be as in **I41**. Then, if $\mathbf{b}$ is positive and non-zero,

(i) $$\tfrac{1}{2}\mathbf{b} \geqslant |\,\mathbf{x}_p - \tilde{z}_p\,|$$

for all large p. Let $\mathbf{z}$ be the real number containing $\{z\}$. Then, by **I40**,

(ii) $$\tfrac{1}{2}\mathbf{b} \geqslant |\,\tilde{z}_p - \mathbf{z}\,|$$

for all large p. Then $\mathbf{b} \geqslant |\,\mathbf{x}_p - \mathbf{z}\,|$ for all large p.
Therefore, by definition **J3**, $\mathbf{z} = \lim\{\mathbf{x}\}$.
On the other hand, if $\{\mathbf{x}\}$ is a succession and $\lim\{\mathbf{x}\} = \mathbf{a}$, and if $\mathbf{b}$ is positive and non-zero, then

$$\tfrac{1}{2}\mathbf{b} \geqslant |\,\mathbf{x}_p - \mathbf{a}\,| \quad \text{for all large } p, \text{ by definition } \mathbf{J3}.$$

Then $$\mathbf{b} \geqslant |\,\mathbf{x}_p - \mathbf{a}\,| + |\,\mathbf{x}_q - \mathbf{a}\,| \quad \text{for all large } p \text{ and } q$$
$$\geqslant |\,\mathbf{x}_p - \mathbf{x}_q\,|.$$

Therefore $\{\mathbf{x}\}$ is a Cauchy sequence.

5. Theorem: *If $\{\mathbf{x}\}$ and $\{\mathbf{y}\}$ are Cauchy sequences, then*

$$\lim\{\mathbf{x}\} + \lim\{\mathbf{y}\} = \lim\{\mathbf{x} + \mathbf{y}\}$$
$$\lim\{\mathbf{x}\} - \lim\{\mathbf{y}\} = \lim\{\mathbf{x} - \mathbf{y}\}$$

and $$\lim\{\mathbf{x}\}\cdot\lim\{\mathbf{y}\} = \lim\{\mathbf{x}\cdot\mathbf{y}\}.$$

Proof: By definitions **J3** and **I3**, $\lim\{\mathbf{x}\} = \mathbf{a}$ if and only if $\{\mathbf{x}\} \in \tilde{\mathbf{a}}$.

Let $\lim\{\mathbf{x}\} = \mathbf{a}$ and $\lim\{\mathbf{y}\} = \mathbf{c}$, using **J4**. By definition **I6**, $\tilde{\mathbf{a}} + \tilde{\mathbf{c}}$ contains $\{\mathbf{x} + \mathbf{y}\}$, and so $\mathbf{a} + \mathbf{c} = \lim\{\mathbf{x} + \mathbf{y}\}$, that is,

$$\lim\{\mathbf{x}\} + \lim\{\mathbf{y}\} = \lim\{\mathbf{x} + \mathbf{y}\}.$$

The proof of the second equation is similar. The third equation follows similarly from definition **I17**.

6. Theorem: *If* $\mathbf{a}_p \geqslant \mathbf{b}_p$ *for all large* p, *then* $\lim\{\mathbf{a}\} \geqslant \lim\{\mathbf{b}\}$.

Proof: $\mathbf{a}_p - \mathbf{b}_p$ is positive for all large p. Therefore the Cauchy number containing $\{\mathbf{a} - \mathbf{b}\}$ is positive, by definition **I29**. In order-isomorphism **I42**, this Cauchy number corresponds to $\lim\{\mathbf{a} - \mathbf{b}\}$, which is equal to $\lim\{\mathbf{a}\} - \lim\{\mathbf{b}\}$, by **J5**. Therefore $\lim\{\mathbf{a}\} \geqslant \lim\{\mathbf{b}\}$.

7. Definition (of *continuous*): The function θ is *continuous* if, for every Cauchy sequence $\{\mathbf{x}\}$ of real numbers, $\{\theta(\mathbf{x})\}$ is a Cauchy sequence, and $\lim\{\theta(\mathbf{x})\} = \theta(\lim\{\mathbf{x}\})$.

Note: Here $\{\theta(\mathbf{x})\}$ denotes, of course, the succession whose p^{th} element is $\theta(\mathbf{x}_p)$.

8. Theorem: *If* θ *and* ϕ *are continuous, then so are the functions* σ, δ, *and* π *defined by*

a. $$\sigma(\mathbf{x}) = \theta(\mathbf{x}) + \phi(\mathbf{x})$$

b. $$\delta(\mathbf{x}) = \theta(\mathbf{x}) - \phi(\mathbf{x})$$

and

c. $$\pi(\mathbf{x}) = \theta(\mathbf{x})\cdot\phi(\mathbf{x})$$

Proof: $\sigma(\lim\{\mathbf{x}\}) = \theta(\lim\{\mathbf{x}\}) + \phi(\lim\{\mathbf{x}\})$

$$= \lim\{\theta(\mathbf{x})\} + \lim\{\phi(\mathbf{x})\}$$

because θ and ϕ are continuous,

$$= \lim\{\theta(\mathbf{x}) + \phi(\mathbf{x})\} \qquad \text{by } \mathbf{J5}$$
$$= \lim\{\sigma(\mathbf{x})\}$$

And by **I5** $\{\theta(\mathbf{x}) + \phi(\mathbf{x})\}$ is a Cauchy sequence. Therefore σ is continuous. Similarly for the others.

9. Definition (of *polynomial*): The function π defined by

$$\pi(\mathbf{x}) = \sum_{p=0}^{n} \mathbf{a}_p \cdot \mathbf{x}^p$$

is a *polynomial*.

10. Theorem: *Every polynomial is continuous.*

Note: We shall use " the function $\mathbf{x}$ " as an abbreviation for " the function ι defined by $\iota(\mathbf{x}) = \mathbf{x}$ ", " the function $\mathbf{a}$ " for " the function α defined by $\alpha(\mathbf{x}) = \mathbf{a}$ ", and so on.

Proof: Clearly $\mathbf{x}$ is continuous. If $\mathbf{x}^r$ is continuous, then so is $\mathbf{x}^{r+1}$, by **J8c**. Therefore, by induction on r, $\mathbf{x}^r$ is continuous for all positive non-zero integers r. Clearly $\mathbf{a}_r$ is continuous. Therefore, by **J8c**, $\mathbf{a}_r \cdot \mathbf{x}^r$ is continuous.

Therefore if $\sum_{p=0}^{n} \mathbf{a}_p \cdot \mathbf{x}^p$ is continuous, so is $\sum_{p=0}^{n+1} \mathbf{a}_p \cdot \mathbf{x}^p$, by **J8a**.

Therefore, by induction on n, every polynomial is continuous.

11. Theorem: *If* $\{\mathbf{x}\}$ *and* $\{\mathbf{y}\}$ *are Cauchy sequences of real numbers, then* $\lim\{\mathbf{x}\} = \lim\{\mathbf{y}\}$ *if and only if, for each positive non-zero* $\mathbf{b}$, $\mathbf{b} \geqslant | \mathbf{x}_p - \mathbf{y}_p |$ *for all large* p.

Proof: $\lim\{\mathbf{x}\} = \lim\{\mathbf{y}\}$ if and only if $\lim\{\mathbf{x}\} - \lim\{\mathbf{y}\} = 0$, which is true if and only if $\lim\{\mathbf{x} - \mathbf{y}\} = 0$, by **J5**. The theorem now follows from definition **J3**.

12. Theorem: *If* θ *is continuous, if* $\mathbf{b}_1 \geqslant \mathbf{a}_1$, *if* $\theta(\mathbf{b}_1) \geqslant 0$, *and if* $0 \geqslant \theta(\mathbf{a}_1)$, *then there is a* $\mathbf{c}$ *such that* $\theta(\mathbf{c}) = 0$ *and* $\mathbf{b}_1 \geqslant \mathbf{c} \geqslant \mathbf{a}_1$.

Note: Less formally, this theorem could be stated as " If a continuous function is positive at one point and negative at another, then it is zero somewhere between them ".

Proof: We shall define inductively successions $\{\mathbf{a}\}$ and $\{\mathbf{b}\}$ with the following properties: (i) $0 \geqslant \theta(\mathbf{a}_p)$; (ii) $\theta(\mathbf{b}_p) \geqslant 0$; (iii) $\mathbf{a}_{p+1} \geqslant \mathbf{a}_p$; (iv) $\mathbf{b}_p \geqslant \mathbf{b}_{p+1}$; and (v) $\mathbf{b}_p - \mathbf{a}_p = 2^{1-p}(\mathbf{b}_1 - \mathbf{a}_1)$ for all p. Clearly $\mathbf{a}_1$ and $\mathbf{b}_1$ have these properties as far as they apply. Suppose that $\mathbf{a}_1 \ldots \mathbf{a}_q$ and $\mathbf{b}_1 \ldots \mathbf{b}_q$ have been defined and have these properties as far as they apply. Put $\mathbf{a}_{q+1}$ and $\mathbf{b}_{q+1}$ equal to $\mathbf{a}_q$ and $\frac{1}{2}(\mathbf{a}_q + \mathbf{b}_q)$ respectively if $\theta(\frac{1}{2}(\mathbf{a}_q + \mathbf{b}_q)) \geqslant 0$; but to $\frac{1}{2}(\mathbf{a}_q + \mathbf{b}_q)$ and $\mathbf{b}_q$ if not.

Clearly $0 \geqslant \theta(\mathbf{a}_{q+1})$, $\theta(\mathbf{b}_{q+1}) \geqslant 0$, $\mathbf{a}_{q+1} \geqslant \mathbf{a}_q$, and $\mathbf{b}_q \geqslant \mathbf{b}_{q+1}$.

Also $\mathbf{b}_{q+1} - \mathbf{a}_{q+1} = \frac{1}{2}(\mathbf{b}_q - \mathbf{a}_q) = 2^{1-(q+1)}(\mathbf{b}_1 - \mathbf{a}_1)$.

Therefore $\mathbf{a}_{q+1}$ and $\mathbf{b}_{q+1}$ have the desired properties.

By (iii) and (iv), $\mathbf{a}_p \geqslant \mathbf{a}_q$ and $\mathbf{b}_q \geqslant \mathbf{b}_p$ whenever $p \geqslant q$.

$\mathbf{b}_p \geqslant \mathbf{a}_p$ for every p, by (v) and because $\mathbf{b}_1 \geqslant \mathbf{a}_1$. Therefore if $p \geqslant r$ and $q \geqslant r$, we have $\mathbf{b}_r \geqslant \mathbf{b}_p \geqslant \mathbf{a}_p \geqslant \mathbf{a}_r$ and $\mathbf{b}_r \geqslant \mathbf{b}_q \geqslant \mathbf{a}_q \geqslant \mathbf{a}_r$. Then

(vi) $\qquad \mathbf{b}_r - \mathbf{a}_r \geqslant | \mathbf{a}_p - \mathbf{a}_q | \qquad$ whenever $p \geqslant r$ and $q \geqslant r$.

If $\mathbf{d}$ is positive and non-zero, there is, using **H9**, a r for which

$$\begin{aligned} \mathbf{d} &\geqslant 2^{1-r}(\mathbf{b}_1 - \mathbf{a}_1) \\ &= \mathbf{b}_r - \mathbf{a}_r \qquad \text{by (v).} \end{aligned}$$

Therefore, by (vi), $\mathbf{d} \geqslant |\mathbf{a}_p - \mathbf{a}_q|$ whenever $p \geqslant r$ and $q \geqslant r$.
Therefore $\{\mathbf{a}\}$ is a Cauchy sequence. Similarly, $\{\mathbf{b}\}$ is a Cauchy sequence.
Moreover,

$$\mathbf{d} \geqslant \mathbf{b}_p - \mathbf{a}_p \text{ whenever } p \geqslant r$$
$$= |\mathbf{b}_p - \mathbf{a}_p|.$$

Therefore $\lim\{\mathbf{a}\} = \lim\{\mathbf{b}\} = \mathbf{c}$, say, by **J11**.
Therefore $\lim(\theta\{\mathbf{a}\}) = \theta(\mathbf{c}) = \lim(\theta\{\mathbf{b}\})$, because θ is continuous. But, from (i) and (ii) and **J6**, we see that $0 \geqslant \lim \theta(\{\mathbf{a}\})$ and $\lim(\theta\{\mathbf{b}\}) \geqslant 0$.
Therefore $\theta(\mathbf{c}) = 0$.

13. Theorem: *If* $\mathbf{d} \geqslant 0$ *and* r *is a positive non-zero integer, there is a* $\mathbf{c}$ *for which* $\mathbf{c}^r = \mathbf{d}$ *and* $\mathbf{c} \geqslant 0$.

Proof: If $\mathbf{x} = 0$, then $0 \geqslant -\mathbf{d} = \mathbf{x}^r - \mathbf{d}$.
If $\mathbf{x} = 1 + \mathbf{d}$, then

$$\mathbf{x}^r = (1 + \mathbf{d})^{r-1} \cdot (1 + \mathbf{d})$$
$$\geqslant 1 \cdot \mathbf{d}.$$

Therefore $\mathbf{x}^r - \mathbf{d} \geqslant 0$. Therefore, by **J10** and **12**, there is a $\mathbf{c}$ for which $\mathbf{c}^r - \mathbf{d} = 0$ and $1 + \mathbf{d} \geqslant \mathbf{c} \geqslant 0$.

14. Definition (of *root*): If r is a positive non-zero integer and $\mathbf{d} \geqslant 0$, the rth *root* of $\mathbf{d}$ is the positive real number $\mathbf{c}$ for which $\mathbf{c}^r = \mathbf{d}$.

Note: That there is one and only one such number $\mathbf{c}$ follows from **J13** and **H11**.

Notation: The rth root of $\mathbf{d}$ is written $\sqrt[r]{\mathbf{d}}$.
Clearly $\sqrt[r]{0} = 0$, $\sqrt[r]{1} = 1$, and $\sqrt[1]{\mathbf{d}} = \mathbf{d}$.

15. Theorem: *If* $p/q = r/s$, $q > 0$, $s > 0$, *and* $\mathbf{d} \geqslant 0$; *then* $\sqrt[q]{\mathbf{d}^p} = \sqrt[s]{\mathbf{d}^r}$.

Proof: $(\sqrt[q]{\mathbf{d}^p})^{q \cdot s} = ((\sqrt[q]{\mathbf{d}^p})^q)^s = (\mathbf{d}^p)^s = \mathbf{d}^{p \cdot s}$. Similarly $(\sqrt[s]{\mathbf{d}^r})^{q \cdot s} = \mathbf{d}^{q \cdot r}$. But $p \cdot s = q \cdot r$, because $p/q = r/s$. Therefore, by **H11**, $\sqrt[q]{\mathbf{d}^p} = \sqrt[s]{\mathbf{d}^r}$.

16. Definition (of rational *power* of a positive real number): If x is rational and $\mathbf{d}$ is positive and x and $\mathbf{d}$ are not both zero, then $\mathbf{d}^x$ is $\sqrt[q]{\mathbf{d}^p}$ where $x = p/q$, p and q are integers, and $q > 0$.

Note: Each rational number x is equal to some p/q with the desired properties. Moreover, if $x = p/q = r/s$, $p > 0$, and $s > 0$, then

$\sqrt[q]{\mathbf{d}^p} = \sqrt[s]{\mathbf{d}^r}$, by **J15**, and so the number defined as $\mathbf{d}^x$ is the same no matter which p and q are chosen, provided that $x = p/q$. Finally, if $x = p/1$, then $\mathbf{d}^x = \sqrt[1]{\mathbf{d}^p} = \mathbf{d}^p$, and so this definition is compatible with the previous one.

17. Theorem: $\mathbf{x}^u \cdot \mathbf{y}^u = (\mathbf{x} \cdot \mathbf{y})^u$ *if* $\mathbf{x} \geqslant 0$, $\mathbf{y} \geqslant 0$, *and* u *is rational and non-zero.*

Proof: Let u be p/q and $q > 0$. Then

$$\begin{aligned} (\mathbf{x}^u \cdot \mathbf{y}^u)^q &= (\sqrt[q]{\mathbf{x}^p} \cdot \sqrt[q]{\mathbf{y}^p})^q \\ &= (\sqrt[q]{\mathbf{x}^p})^q \cdot (\sqrt[q]{\mathbf{y}^p})^q && \text{by } \mathbf{H4} \\ &= \mathbf{x}^p \cdot \mathbf{y}^p \\ &= (\mathbf{x} \cdot \mathbf{y})^p && \text{by } \mathbf{H4} \\ &= (\sqrt[q]{(\mathbf{x} \cdot \mathbf{y})^p})^q. \end{aligned}$$

Therefore

$$\begin{aligned} \mathbf{x}^u \cdot \mathbf{y}^u &= \sqrt[q]{(\mathbf{x} \cdot \mathbf{y})^p} && \text{by } \mathbf{H11} \\ &= (\mathbf{x} \cdot \mathbf{y})^{p/q} \\ &= (\mathbf{x} \cdot \mathbf{y})^u. \end{aligned}$$

18. Theorem: *If* $\mathbf{x} \geqslant 0$ *and if* u *and* v *are rational and non-zero, then* $(\mathbf{x}^u)^v = \mathbf{x}^{u \cdot v}$.

Proof: Let u be p/q and v be r/s, where $q > 0$ and $s > 0$. $((\mathbf{x}^u)^v)^{q \cdot s} = ((\sqrt[s]{(\mathbf{x}^u)^r})^s)^q = (\mathbf{x}^u)^{r \cdot q} = (\sqrt[q]{\mathbf{x}^p})^{r \cdot q} = \mathbf{x}^{p \cdot r}$. Therefore $(\mathbf{x}^u)^v = \sqrt[q \cdot s]{\mathbf{x}^{p \cdot r}} = \mathbf{x}^{p \cdot r/q \cdot s} = \mathbf{x}^{u \cdot v}$.

19. Theorem: *If* $\mathbf{x} \geqslant 0$ *and if* u *and* v *are rational and non-zero, then* $\mathbf{x}^u \cdot \mathbf{x}^v = \mathbf{x}^{u+v}$.

Proof: Let u be p/q and v be r/s, where $q > 0$ and $s > 0$. Then $u = p \cdot s/q \cdot s$ and $v = q \cdot r/q \cdot s$. Therefore

$$\begin{aligned} \mathbf{x}^u \cdot \mathbf{x}^v &= (\mathbf{x}^{p \cdot s})^{1/q \cdot s} \cdot (\mathbf{x}^{q \cdot r})^{1/q \cdot s} && \text{by } \mathbf{J18} \\ &= (\mathbf{x}^{p \cdot s} \cdot \mathbf{x}^{q \cdot r})^{1/q \cdot s} && \text{by } \mathbf{J17} \\ &= (\mathbf{x}^{p \cdot s + q \cdot r})^{1/q \cdot s} && \text{by } \mathbf{H6} \\ &= \mathbf{x}^{(p \cdot s + q \cdot r)/q \cdot s} && \text{by } \mathbf{J18} \\ &= \mathbf{x}^{u+v}. \end{aligned}$$

20. Lemma (to **J21** and **J23**): *If* $\mathbf{x} > \mathbf{y} > 0$ *and* $q > 0$, *then* $\mathbf{x}^{1/q} > \mathbf{y}^{1/q}$.

Proof: If $\mathbf{y}^{1/q} \geqslant \mathbf{x}^{1/q}$, then

$$\begin{aligned} \mathbf{y} &= (\mathbf{y}^{1/q})^q \\ &\geqslant (\mathbf{x}^{1/q})^q && \text{by } \mathbf{H10} \\ &= \mathbf{x}. \end{aligned}$$

21. Theorem: *If* $\mathbf{x} > 1$, *if* $u > v$, *and if* u *and* v *are rational, then* $\mathbf{x}^u > \mathbf{x}^v$.

Proof: Let u be p/q and v be r/s, where $q > 0$ and $s > 0$. Then $p \cdot s > q \cdot r$. Therefore $\mathbf{x}^{p \cdot s} > \mathbf{x}^{q \cdot r}$, by **H12**.

Therefore $\mathbf{x}^u = \mathbf{x}^{p \cdot s/q \cdot s} > \mathbf{x}^{q \cdot r/q \cdot s} = \mathbf{x}^v$, by **J20**.

Note: It follows that if $1 > \mathbf{y} > 0$, then $\mathbf{y}^v > \mathbf{y}^u$, by applying the theorem to $\mathbf{y}^{-1}$.

22. Lemma (to **J23** and **26**): *If* $\mathbf{d} \geqslant 0$ *and* r *is a positive integer, then* $\mathbf{d}^r \geqslant 1 + r \cdot (\mathbf{d} - 1)$.

Proof: $\mathbf{d}^1 \geqslant 1 + (\mathbf{d} - 1)$.

If $\mathbf{d}^r \geqslant 1 + r \cdot (\mathbf{d} - 1)$, then

$$\begin{aligned} \mathbf{d}^{r+1} &= \mathbf{d}^r \cdot \mathbf{d} \\ &\geqslant (1 + r \cdot (\mathbf{d} - 1)) \cdot (1 + \mathbf{d} - 1) \\ &= 1 + (r + 1) \cdot (\mathbf{d} - 1) + r \cdot (\mathbf{d} - 1)^2 \\ &\geqslant 1 + (r + 1) \cdot (\mathbf{d} - 1). \end{aligned}$$

Therefore, by induction, the lemma is true.

23. Lemma (to **J24**): *If* $\mathbf{a} > 1$ *and* $\mathbf{b} > 0$, *then there is an integer* r *such that* $\mathbf{b} \geqslant \mathbf{a}^{1/r} - 1$ *and* $r > 0$.

Proof: By **J2**, there is an integer r such that

(i) $$r \geqslant (\mathbf{a} - 1)/\mathbf{b} > 0.$$

Now $$\mathbf{a} = (\mathbf{a}^{1/r})^r \geqslant 1 + r \cdot (\mathbf{a}^{1/r} - 1) \qquad \text{by } \mathbf{J22}.$$

Therefore $$(\mathbf{a} - 1)/r \geqslant \mathbf{a}^{1/r} - 1.$$

Then, by (i), $$\mathbf{b} \geqslant (\mathbf{a} - 1)/r \geqslant \mathbf{a}^{1/r} - 1.$$

24. Lemma (to definition **J25**): *If* $\{u\}$ *and* $\{v\}$ *are Cauchy sequences of rational numbers, if* $\lim\{u\} = \lim\{v\}$, *if* $\mathbf{a} > 0$, *and if* $\mathbf{x}_p = \mathbf{a}^{u_p}$ *and* $\mathbf{y}_p = \mathbf{a}^{v_p}$, *then* $\{\mathbf{x}\}$ *and* $\{\mathbf{y}\}$ *are Cauchy sequences and* $\lim\{\mathbf{x}\} = \lim\{\mathbf{y}\}$.

Proof that $\{\mathbf{x}\}$ *is a Cauchy sequence*:

Case I, $\mathbf{a} > 1$. By **I15**, there is a k such that $k \geqslant u_q$ for all large q. Then $\mathbf{a}^k \geqslant \mathbf{a}^{u_q}$. Therefore, for all large p and q,

$$\mathbf{a}^k \cdot (\mathbf{a}^{|u_p - u_q|} - 1) \geqslant |\, \mathbf{a}^{u_p} - \mathbf{a}^{u_q} \,| = |\, \mathbf{x}_p - \mathbf{x}_q \,|.$$

If $\mathbf{b}$ is positive and non-zero so is $\mathbf{b} \cdot \mathbf{a}^{-k}$; therefore there is, by **J23**, a r such that $r > 0$ and $\mathbf{b} \cdot \mathbf{a}^{-k} \geqslant \mathbf{a}^{1/r} - 1$.

Now $1/r \geqslant |u_p - u_q|$ for all large p and q.
Then $\mathbf{b} \geqslant \mathbf{a}^{k}\cdot(\mathbf{a}^{1/r} - 1) \geqslant \mathbf{a}^{k}\cdot(\mathbf{a}^{|u_p-u_q|} - 1) \geqslant |\mathbf{x}_p - \mathbf{x}_q|$.
Therefore $\{\mathbf{x}\}$ is a Cauchy sequence.

Case II, $\mathbf{a} = 1$. Then $\{\mathbf{x}\} = [1]$, which is a Cauchy sequence by **I10**.

Case III, $1 > \mathbf{a} > 0$. Then $\mathbf{a}^{-1} > 1$. $\mathbf{x}_p^{-1} = (\mathbf{a}^{u_p})^{-1} = (\mathbf{a}^{-1})^{u_p}$.

Therefore, by case I, $\{\mathbf{x}^{-1}\}$ is a Cauchy sequence. $\mathbf{x}_p^{-1} \neq 0$.
Therefore, by **I25**, $\{\mathbf{x}\}$ is a Cauchy sequence.

Similarly, $\{\mathbf{y}\}$ is a Cauchy sequence.

Proof that $\lim\{\mathbf{x}\} = \lim\{\mathbf{y}\}$:

Case I, $\mathbf{a} > 1$. By **I15**, there is a k such that $k \geqslant u_p$ and $k \geqslant v_p$ for all large p. Then

$$\mathbf{a}^{k}\cdot(\mathbf{a}^{|u_p-v_p|} - 1) \geqslant |\mathbf{a}^{u_p} - \mathbf{a}^{v_p}| = |\mathbf{x}_p - \mathbf{y}_p|.$$

If $\mathbf{b}$ is positive and non-zero, there is, by **J23**, a r such that

$$\mathbf{b}\cdot\mathbf{a}^{-k} \geqslant \mathbf{a}^{1/r} - 1.$$

By **J11** $\quad 1/r \geqslant |u_p - v_p|$ for all large p.

Then

$$\begin{aligned} \mathbf{b} &\geqslant \mathbf{a}^{k}\cdot(\mathbf{a}^{1/r} - 1) \\ &\geqslant \mathbf{a}^{k}\cdot(\mathbf{a}^{|u_p-v_p|} - 1) \\ &\geqslant |\mathbf{x}_p - \mathbf{y}_p|. \end{aligned}$$

Therefore $\lim\{\mathbf{x}\} = \lim\{\mathbf{y}\}$, by **J11**.

Case II, $a = 1$. $\{\mathbf{x}\} = \{\mathbf{y}\} = [1]$, and so $\lim\{\mathbf{x}\} = \lim\{\mathbf{y}\} = 1$.

Case III, $1 > \mathbf{a} > 0$. Then $\mathbf{a}^{-1} > 1$. Therefore, by case I,

$$\lim\{\mathbf{x}^{-1}\} = \lim\{\mathbf{y}^{-1}\}.$$

But $\lim\{\mathbf{x}^{-1}\}\cdot\lim\{\mathbf{x}\} = \lim[1] = \lim\{\mathbf{y}^{-1}\}\cdot\lim\{\mathbf{y}\}$, by **J5**.
Therefore $\quad \lim\{\mathbf{x}\} = \lim\{\mathbf{y}\}$.

25. Definition (of *power* of a positive real number): If $\mathbf{x} > 0$ and $\mathbf{y} = \lim\{u\}$, then $\mathbf{x}^{\mathbf{y}}$ is $\lim\{\mathbf{x}^{u}\}$. *Note*: by **J24**, $\{\mathbf{x}^{u}\}$ is a Cauchy sequence and therefore has a limit; and this limit is independent of the particular $\{u\}$ chosen, as long as $\lim\{u\} = \mathbf{y}$. If $\mathbf{y}$ is rational, then $\mathbf{x}^{\mathbf{y}}$ is already defined. But under the new definition,

$$\begin{aligned} \mathbf{x}^{\mathbf{y}} &= \lim\{\mathbf{x}^{u}\} && \text{where } \{u\} = [\mathbf{y}] \\ &= \lim[\mathbf{x}^{\mathbf{y}}] && \text{where } \mathbf{x}^{\mathbf{y}} \text{ is as previously defined} \\ &= \mathbf{x}^{\mathbf{y}} && \text{as previously defined.} \end{aligned}$$

Finally, we define $0^{\mathbf{y}}$ to be 0 for every non-zero $\mathbf{y}$.

heorem: *If* $\{\mathbf{x}\}$ *and* $\{v\}$ *are Cauchy sequences, if every* v_p *is* r…, *if every* $\mathbf{x}_p > 0$, *if* $\lim\{\mathbf{x}\} = \mathbf{a} > 0$, *and if* $\lim\{v\} = \mathbf{v}$, *then* $\lim\{\mathbf{x}^v\} = \mathbf{a}^{\mathbf{v}}$.

Notes: $\{\mathbf{x}^v\}$ denotes here, of course, the succession whose p^{th} member is $\mathbf{x}_p{}^{v_p}$. The theorem could be stated less formally as "$\mathbf{x}^{\mathbf{y}}$ is a continuous function of $\mathbf{x}$ and of $\mathbf{y}$ if $\mathbf{x} > 0$".

Proof: By **I15** and **J2**, there is a positive non-zero integer s such that $s \geqslant v_p$ and $s \geqslant -v_p$ for every p. Therefore there is a $\mathbf{k}$ such that $\mathbf{k} \geqslant |\mathbf{a}^{v_p}|$ for every p, and $\mathbf{k}$ is positive and non-zero.

If $\mathbf{f}$ is positive and non-zero, then so is $\mathbf{a}\cdot\mathbf{f}$. Then, because $\mathbf{a} = \lim\{\mathbf{x}\}$, $\mathbf{a}\cdot\mathbf{f} \geqslant |\mathbf{x}_p - \mathbf{a}|$ for all large p; that is, $\mathbf{f} \geqslant |\mathbf{h}|$ where $\mathbf{h}$ is $\mathbf{x}_p/\mathbf{a} - 1$. Thus, if $\mathbf{b}$ is positive and non-zero, then $1 \geqslant |\mathbf{h}|$, $\frac{1}{4}\mathbf{b}/(\mathbf{k}\cdot s) \geqslant |\mathbf{h}|$, $1/2\cdot s \geqslant |\mathbf{h}|$, and $\mathbf{b}/\mathbf{k}\cdot 2^{s+1} \geqslant |\mathbf{h}|$ for all large p.

If $0 > \mathbf{h}$ and $v_p \geqslant 0$, then

$$\begin{aligned} 0 \geqslant (1+\mathbf{h})^{v_p} - 1 &\geqslant (1+\mathbf{h})^s - 1 && \text{by note } \mathbf{J21} \\ &\geqslant s\cdot\mathbf{h} && \text{by } \mathbf{J22}\text{, with } 1+\mathbf{h} \text{ for } \mathbf{d} \\ &\geqslant -\tfrac{1}{2}\mathbf{b}/\mathbf{k}. \end{aligned}$$

If $0 > \mathbf{h}$ and $0 > v_p$, then

$$\begin{aligned} (1+\mathbf{h})^s &\geqslant 1 + s\cdot\mathbf{h} && \text{by } \mathbf{J22} \text{ with } 1+\mathbf{h} \text{ for } \mathbf{d} \\ &\geqslant (1 - 2\cdot s\cdot\mathbf{h})^{-1} && \text{because } (1-2\cdot s\cdot\mathbf{h})(1+s\cdot\mathbf{h}) \geqslant 1.^* \end{aligned}$$

Therefore $(1 - 2\cdot s\cdot\mathbf{h}) \geqslant (1+\mathbf{h})^{-s}$.

Therefore
$$\begin{aligned} \tfrac{1}{2}\mathbf{b}/\mathbf{k} &\geqslant -2\cdot s\cdot\mathbf{h} \\ &\geqslant (1+\mathbf{h})^{-s} - 1 \\ &\geqslant (1+\mathbf{h})^{v_p} - 1 \\ &\geqslant 0. \end{aligned}$$

If $\mathbf{h} \geqslant 0$ and $0 > v_p$, then

$$\text{(i)} \quad \begin{cases} \tfrac{1}{2}\mathbf{b}/\mathbf{k} \geqslant (1+\mathbf{h})^s - 1 & \text{(see the footnote on the next page)} \\ \qquad \geqslant 1 - (1+\mathbf{h})^{-s} \end{cases}$$

Then
$$\begin{aligned} 0 &\geqslant (1+\mathbf{h})^{v_p} - 1 \\ &\geqslant (1+\mathbf{h})^{-s} - 1 \\ &\geqslant -\tfrac{1}{2}\mathbf{b}/\mathbf{k} && \text{by (i).} \end{aligned}$$

* *Proof*: $1 \geqslant -2\cdot s\cdot\mathbf{h}$. Therefore $-s\cdot\mathbf{h} \geqslant 2\cdot s^2\cdot\mathbf{h}^2$. Therefore $1 - s\cdot\mathbf{h} - 2\cdot s^2\cdot\mathbf{h}^2 \geqslant 1$.

If $\mathbf{h} \geqslant 0$ and $v_p \geqslant 0$, then

$$\begin{aligned} \tfrac{1}{2}\mathbf{b}/\mathbf{k} &\geqslant 2^s\cdot\mathbf{h} \\ &\geqslant (1+\mathbf{h})^s - 1 && \text{(this is readily proved by induction on } s)^* \\ &\geqslant (1+\mathbf{h})^{v_p} - 1 && \text{by } \mathbf{J21} \\ &\geqslant 0. \end{aligned}$$

Therefore, in all four cases,

$$\begin{aligned} \tfrac{1}{2}\mathbf{b}/\mathbf{k} &\geqslant |\,(1+\mathbf{h})^{v_p} - 1\,| \\ &= |\,(\mathbf{x}_p/\mathbf{a})^{v_p} - 1\,|. \end{aligned}$$

Therefore $\tfrac{1}{2}\mathbf{b} \geqslant \tfrac{1}{2}\mathbf{b}\cdot|\,\mathbf{a}^{v_p}\,|/\mathbf{k} \geqslant |\,\mathbf{x}_p{}^{v_p} - \mathbf{a}^{v_p}\,|$.
But $\tfrac{1}{2}\mathbf{b} \geqslant |\,\mathbf{a}^{v_p} - \mathbf{a}^{\mathbf{v}}\,|$ for all large p, because $\mathbf{a}^{\mathbf{v}} = \lim\{\mathbf{a}^v\}$.
Therefore $\mathbf{b} \geqslant |\,\mathbf{x}_p{}^{v_p} - \mathbf{a}^{\mathbf{v}}\,|$ for all large p. Therefore $\lim\{\mathbf{x}^v\} = \mathbf{a}^{\mathbf{v}}$.

Corollary: Put $\mathbf{x}^{u_p}$ in the place of $\mathbf{x}_p$, where $\mathbf{x} > 0$ and $\{u\}$ is a Cauchy sequence of rational numbers. Let $\lim\{u\} = \mathbf{u}$. Then in place of $\mathbf{a}$ we have $\mathbf{x}^{\mathbf{u}}$, and the theorem becomes

$$\lim\{(\mathbf{x}^u)^v\} = (\mathbf{x}^{\mathbf{u}})^{\mathbf{v}}.$$

27. Theorem: *If* $\mathbf{x} > 0$ *and* $\mathbf{y} > 0$, *then* $\mathbf{x}^{\mathbf{u}}\cdot\mathbf{y}^{\mathbf{u}} = (\mathbf{x}\cdot\mathbf{y})^{\mathbf{u}}$, $(\mathbf{x}^{\mathbf{u}})^{\mathbf{v}} = \mathbf{x}^{\mathbf{u}\cdot\mathbf{v}}$, *and* $\mathbf{x}^{\mathbf{u}}\cdot\mathbf{x}^{\mathbf{v}} = \mathbf{x}^{\mathbf{u}+\mathbf{v}}$.

Proof: Let $\mathbf{u}$ be $\lim\{u\}$ and $\mathbf{v}$ be $\lim\{v\}$. Then

$$\begin{aligned} \mathbf{x}^{\mathbf{u}}\cdot\mathbf{y}^{\mathbf{u}} &= \lim\{\mathbf{x}^u\}\cdot\lim\{\mathbf{y}^u\} \\ &= \lim\{\mathbf{x}^u\cdot\mathbf{y}^u\} && \text{by } \mathbf{J5} \\ &= \lim\{(\mathbf{x}\cdot\mathbf{y})^u\} && \text{by } \mathbf{J17} \\ &= (\mathbf{x}\cdot\mathbf{y})^{\mathbf{u}}. \end{aligned}$$

And
$$\begin{aligned} (\mathbf{x}^{\mathbf{u}})^{\mathbf{v}} &= \lim\{(\mathbf{x}^u)^v\} && \text{by corollary } \mathbf{J26} \\ &= \lim\{\mathbf{x}^{u\cdot v}\} && \text{by } \mathbf{J18} \\ &= \mathbf{x}^{\mathbf{u}\cdot\mathbf{v}} && \text{because } \lim\{u\cdot v\} = \mathbf{u}\cdot\mathbf{v}. \end{aligned}$$

And
$$\begin{aligned} \mathbf{x}^{\mathbf{u}}\cdot\mathbf{x}^{\mathbf{v}} &= \lim\{\mathbf{x}^u\}\cdot\lim\{\mathbf{x}^v\} \\ &= \lim\{\mathbf{x}^u\cdot\mathbf{x}^v\} && \text{by } \mathbf{J5} \\ &= \lim\{\mathbf{x}^{u+v}\} && \text{by } \mathbf{J19} \\ &= \mathbf{x}^{\mathbf{u}+\mathbf{v}} && \text{because } \lim\{u+v\} = \mathbf{u}+\mathbf{v}. \end{aligned}$$

* Assume that $2^s\cdot\mathbf{h} \geqslant (1+\mathbf{h})^s - 1 + \mathbf{h}$ and multiply by $1+\mathbf{h}$. It soon follows that $2^{s+1}\cdot\mathbf{h} \geqslant (1+\mathbf{h})^{s+1} - 1 + \mathbf{h}$.

28. Notation: $[\mathbf{x}]$ denotes the integer p for which $p + 1 > \mathbf{x} \geqslant p$. (This notation should not be confused with notation **I10**.)

Clearly $[\mathbf{x}]$ is defined for all real numbers $\mathbf{x}$; and $\mathbf{x} \geqslant [\mathbf{x}] > \mathbf{x} - 1$.

Theorem: *If t is an integer, $t \geqslant 2$, and $1 > \mathbf{x} \geqslant 0$, then there is a succession $\{r\}$ of integers such that $t > r_p \geqslant 0$ for every p, and* $\lim \{c\} = \mathbf{x}$, *where c_p is $\sum_{m=1}^{p} r_m \cdot t^{-m}$.*

Proof: Let r_p be $[\mathbf{x} \cdot t^p] - t \cdot [\mathbf{x} \cdot t^{p-1}]$. Then each r_p is an integer.

$$c_p = \sum_{m=1}^{p} ([\mathbf{x} \cdot t^m]/t^m - [\mathbf{x} \cdot t^{m-1}]/t^{m-1}) = [\mathbf{x} \cdot t^p]/t^p.$$

Therefore $\quad \mathbf{x} - c_p \geqslant \mathbf{x} - \mathbf{x} \cdot t^p/t^p = 0.$

And $\quad c_p - \mathbf{x} \geqslant (\mathbf{x} \cdot t^p - 1)/t^p - \mathbf{x} = -t^{-p}.$

Therefore $\quad t^{-p} \geqslant | c_p - \mathbf{x} |.$

If $\mathbf{b} > 0$, there is a positive q such that $2^q \geqslant \mathbf{b}^{-1}$, by **H9** and **J2**. Then $\mathbf{b} \geqslant 2^{-q} \geqslant t^{-q}$, because $t \geqslant 2$. And $t^{-q} \geqslant t^{-p}$ whenever $p \geqslant q$. Therefore $\lim \{c\} = \mathbf{x}$, by definition **J3**.

$$r_p > (\mathbf{x} \cdot t^p - 1) - t \cdot \mathbf{x} \cdot t^{p-1} = -1$$

and $\quad t = \mathbf{x} \cdot t^p - t \cdot (\mathbf{x} \cdot t^{p-1} - 1) > r_p.$

Therefore $\quad t > r_p \geqslant 0.$

Note: If t is *ten*, then the succession $\{c\}$ corresponds to an infinite decimal, because its pth element is

$$r_1/10 + r_2/10^2 + \ldots + r_p/10^p$$

that is, $0 \cdot r_1 r_2 r_3 \ldots r_p$ in the usual decimal notation. Its limit is, therefore, what is usually meant by the infinite decimal $0 \cdot r_1 r_2 r_3 \ldots$. The theorem in this case becomes " Every number between 0 and 1 can be expressed as an infinite decimal ". It follows that every real number can be so represented, because $\mathbf{x} = [\mathbf{x}] + \mathbf{r}$, where $1 > \mathbf{r} \geqslant 0$. $[\mathbf{x}]$ is an integer, and so has the usual decimal representation, $n_1 n_2 \ldots n_q$, say, where each n is an integer between 0 and 9 inclusive. $\mathbf{r}$ can be expressed in the form $0 \cdot r_1 r_2 r_3 \ldots$ and then $\mathbf{x}$ is $n_1 n_2 \ldots n_q \cdot r_1 r_2 r_3 \ldots$.

It is obvious that each infinite decimal is the limit of a Cauchy sequence of rational numbers. Therefore the real numbers are precisely the infinite decimals.

EXERCISES J

1*. Let F be the set of all expressions of the form $P(x)/Q(x)$, where P and Q are polynomials with non-zero leading coefficients, together with 0. Addition and multiplication are defined as in elementary algebra. Prove that F is a field and that if we define $(ax^n + bx^{n-1} + \ldots + d)/(fx^m + gx^{m-1} + \ldots + k)$ to be positive if and only if $a \cdot f$ is positive, then F is an ordered field. Prove that theorem **J2** is not true for F (that is, if ξ is an element of F, there is not always an integer r such that $r \geqslant \xi$).

2. k is an *upper bound* of a set M if $k \geqslant m$ whenever $m \in M$. Prove that if M is a set of real numbers which has an upper bound, then there is an upper bound l with the property that if k is any upper bound, then $k \geqslant l$.

3. P and Q are non-null sets of real numbers such that each real number is in either P or Q, no real number is in both, and if $p \in P$ and $q \in Q$ then $p \geqslant q$. Prove that there is a real number k with the property that $x \in P$ if $x > k$, and $x \in Q$ if $k > x$. Prove that this is not true if we replace " real " by " rational " throughout. *Note*: this is *Dedekind's theorem.*

4*. The decimal $\{c\}$ (see Note **J28**) is said to *terminate* if $c_p = 0$ for all large p. Prove that two different non-terminating decimals must represent two unequal numbers; that two different terminating decimals must represent two unequal numbers; and that each non-zero terminating decimal represents the same number as just one non-terminating decimal [e.g. $0{\cdot}5 = 0{\cdot}4\dot{9}$].

CHAPTER K

Complex Numbers

1. Definition (of *complex* number): A *complex* number is an ordered pair of real numbers, sums and products being defined by the equations:

$$(\mathbf{a}, \mathbf{b}) + (\mathbf{c}, \mathbf{d}) = (\mathbf{a} + \mathbf{c}, \mathbf{b} + \mathbf{d})$$

and

$$(\mathbf{a}, \mathbf{b})\cdot(\mathbf{c}, \mathbf{d}) = (\mathbf{a}\cdot\mathbf{c} - \mathbf{b}\cdot\mathbf{d}, \mathbf{a}\cdot\mathbf{d} + \mathbf{b}\cdot\mathbf{c})$$

Note: The ordered pair $(\mathbf{a}, \mathbf{b})$ is equal to the ordered pair $(\mathbf{c}, \mathbf{d})$ if and only if $\mathbf{a} = \mathbf{c}$ and $\mathbf{b} = \mathbf{d}$.

2. Theorem: *The complex numbers form a field.*

Proof: **D1a, b, c, d, e, f, g,** and **k** can be verified immediately from definitions **K1** and the properties of real numbers, if we take $(0, 0)$ for o, $(1, 0)$ for i, and $(-\mathbf{a}, -\mathbf{b})$ for $-(\mathbf{a}, \mathbf{b})$.

To verify **D1h**, we have first to show that if $(\mathbf{a}, \mathbf{b}) \neq (0, 0)$, then $\mathbf{a}^2 + \mathbf{b}^2 \neq 0$. In fact, $\mathbf{a}^2$ and $\mathbf{b}^2$ are positive, by **F3**. If $\mathbf{a}^2 + \mathbf{b}^2 = 0$, then $-\mathbf{a}^2 = \mathbf{b}^2$ and so is positive. Therefore $\mathbf{a}^2 = 0$, and so $\mathbf{a} = 0$. Then $\mathbf{b} = 0$, contradicting the condition $(\mathbf{a}, \mathbf{b}) \neq (0, 0)$. Therefore $\mathbf{a}^2 + \mathbf{b}^2$ cannot be zero, and so we can put $(\mathbf{a}, \mathbf{b})^- = (\mathbf{a}\cdot\mathbf{c}, -\mathbf{b}\cdot\mathbf{c})$, where $\mathbf{c} = 1/(\mathbf{a}^2 + \mathbf{b}^2)$. It is then easy to verify **D1h.**

3. Definition (of $\mathbf{i}$): $\mathbf{i}$ is $(0, 1)$. Note that $\mathbf{i}$ is not the complex number playing the part of the i defined in **D1**. This, as mentioned in **K2**, is the complex number $(1, 0)$.

$$\mathbf{i}^2 = (0\cdot0 - 1\cdot1, 0\cdot1 + 1\cdot0) = (-1, 0) = -(1, 0)$$
$$(\mathbf{a}, \mathbf{b}) = (\mathbf{a}, 0) + (0, \mathbf{b}) = (\mathbf{a}, 0) + (0, 1)(\mathbf{b}, 0)$$

4. Theorem: $(\mathbf{x}, 0) \longleftrightarrow \mathbf{x}$ *is an isomorphism between complex numbers of the form* $(\mathbf{x}, 0)$ *and the field of real numbers.*

Proof: The correspondence is clearly a matching.

$$(\mathbf{x}, 0) + (\mathbf{y}, 0) = (\mathbf{x} + \mathbf{y}, 0)$$
$$(\mathbf{x}, 0)\cdot(\mathbf{y}, 0) = (\mathbf{x}\cdot\mathbf{y} - 0\cdot0, \mathbf{x}\cdot0 + 0\cdot\mathbf{y}) = (\mathbf{x}\cdot\mathbf{y}, 0\cdot0)$$

5. We call complex numbers of the form (**x**, 0) *real.* If we ignore the distinction between a real complex number and the real number corresponding to it in isomorphism **K4**, it will follow from **K3** that each complex number is of the form $\mathbf{x} + \mathbf{i}\cdot\mathbf{y}$, where **x** and **y** are real numbers and $\mathbf{i}^2 = -1$. Moreover the definitions of multiplication and addition are what would be obtained by writing (**a**, **b**) as $\mathbf{a} + \mathbf{i}\cdot\mathbf{b}$ and so on, and multiplying out as though **i** were a real number, and then replacing $\mathbf{i}^2$ by -1 wherever it occurs. It follows that every calculation with complex numbers can be carried out by this process, and that we can now treat the complex numbers from the usual elementary point of view.

BIBLIOGRAPHY

A. A. ALBERT: *Modern Higher Algebra*, University of Chicago Press, 1937.

G. BIRKHOFF and S. MACLANE: *A Brief Survey of Modern Algebra*, Macmillan, 1941 (2nd ed., 1965).

R. DEDEKIND: *Was sind und was sollen die Zahlen?* (*Gesammelte Werke*, Vol. III, p. 335), Vieweg, 1930. English trans. in *Essays on the Theory of Numbers* by R. Dedekind, Dover reprint, 1963.

G. H. HARDY: *Pure Mathematics*, Cambridge University Press, 1908 (11th ed., 1959).

E. LANDAU: *Grundlagen der Analysis*, Akademische Verlagsgesellschaft M.B.H., 1930. English trans., *Foundations of Analysis*, Chelsea Pub. Co., 1951.

J. E. LITTLEWOOD: *The Elements of the Theory of Real Functions*, Heffer, 1936. Third rev. ed., Dover Publications, Inc., 1954.

C. C. MACDUFFEE: *An Introduction to Abstract Algebra*, Wiley, 1947. Reprint of 5th (1956) ed., Dover Publications, Inc., 1966.

B. RUSSELL: *Introduction to Mathematical Philosophy*, Allen and Unwin, 1919. Reprinted by Humanities Press, Inc.

A. N. WHITEHEAD and B. RUSSELL: *Principia Mathematica*, Cambridge University Press, 1910 (2nd ed., 3 vols., 1925–27).

KEY TO THE EXERCISES

In general these are skeleton solutions and hints, from which the full solutions should be easy to deduce.

A

1. $0'' + 0'' = (0'' + 0')' = ((0'' + 0)')' = ((0'')')' = 0''''$, by **3a** and **b**.

$$\begin{aligned} 0''\cdot 0'' &= 0''\cdot 0' + 0'' && \text{by } \mathbf{9a} \\ &= 0'' + 0'' && \text{by } \mathbf{14} \text{ and } \mathbf{10} \\ &= 0'''' && \text{(already proved).} \end{aligned}$$

2. Let M be the set of x for which $x \neq x'$. By **1b**, $0 \in M$. **1a** shows that if $x \neq x'$, then $x' \neq x''$. Then **1c** completes the proof.

3. $0 + y = x + y$ by **3a**. Then $x = 0$ by **6**.

4. $y = (y + u) + v = y + (u + v)$, by **4**. Then $u + v = 0$ by **10** and exercise 3.

Then $u = 0$, by **7**.

5. $x + (u + v) = (x + u) + v = y + v = x' = x + 0'$, by **4**, **3a** and **3b**. Then $u + v = 0'$. If $v \neq 0$, then $v = w'$ by **2**; and then $(u + w)' = u + w' = u + v = 0'$. Then $u + w = 0$ and so $u = 0$.

6. Let N be the set of all whole numbers p with the following property: if $y \in X$, then there is a u such that $y = p + u$. Now $m' + u = m + u' \neq m$, by **3b**, exercise 3, and **2**. Therefore if $m \in X$, $m' \notin N$. Thus not every whole number is in N. Now clearly $0 \in N$; therefore there must be a x such that $x \in N$ and $x' \notin N$. We shall show that this is the x required. Since $x \in N$ it has the property stated. It remains to show that $x \in X$. $x' \notin N$; therefore there is a m in X such that for *no* v can we have $m = x' + v$. Since $m \in X$, however, $m = x + w$ for some w. If $w \neq 0$, then $w = v'$ for some v, and so $m = x + v' = x' + v$ which we have just seen not to be so. Therefore $x = m \in X$.

7. Let X be the set of all whole numbers v for which $b\cdot v = a + t$ for some non-zero t; such numbers clearly exist, e.g. a' is one. By

exercise 6, there is a x in X such that if $y \in X$ then $y = x + u$ for some u. Clearly $x \neq 0$, and so $x = q'$ for some q. By **8** we can now prove that $a = b \cdot q + r$ for some r, and so (i) is true.

If (ii) is *not* true, then $r = b + w$ by 8. Then $a = b \cdot q + (b + w) = b \cdot q' + w$, which is not so, because $q' \in X$.

8. **2**, **3**, **4**, **5**, **8**, **9**, **10**, **12**, **13**, **14**, and **15** are true in all cases, as only **1c** is required in their proof.

In case (i), **6** and **16** are also true, **1b** not being required. **7** and **11** may be true or false if **1b** is false. E.g. in the system $\{0, 0', 0'', 0'''\}$ with $0'''' = 0$, **1a** and **1c** are true and **1b** is false. $0' + 0''' = 0$, and so **7** is false. $0'' \cdot 0'' = 0$, and so **11** is false. But in the system $\{0\}$ with $0' = 0$, **7** and **11** are clearly true, although **1b** is false (and **1a** and **1c** are true).

In case (ii), **7** and **11** are true, **1a** not being required for their proofs. If **1a** is false, then $u' = v'$ where $u \neq v$. Then $u + 0' = v + 0'$, but $u \neq v$. That is, **6** is false.

The general form of sequence for which **1b** and **1c** are true and **1a** is false is $x_0, x_1, x_2, \ldots . x_r, y_1, y_2, \ldots , y_s$ where $y_s' = y_1$. Denote by $*$ the addition which **3** defines in this sequence. Then $y_i * y_i$ turns out to be y_k where $k \equiv r + 2i$ modulo s. If we choose i congruent to $-r$ modulo s, then $0'' \cdot y_i = y_k$ where $k \equiv r - 2r \equiv -r \equiv i$. Then, using **10**, $0'' \cdot y_i = 0' \cdot y_i$, and so **16** is false.

B

1. (i) A hemigroup except for the existence of e.
   (ii) A group.
   (iii) $*$ is not an operation unless m is prime. In this case we have a group.
   (iv) $1 * x = 1$ for every x. Therefore **c** fails.
   (v) If $x * y = 1$ then $(x - 1) \cdot (y - 1) = 0$. Thus $*$ is an operation on the given set. **a**, **b**, and **c** are easily proved, and $0 * 0 = 0$, which proves **d**. Therefore we have a hemigroup. $3 + x - 3 \cdot x = 0$ is not solvable for x. Therefore the hemigroup is not a group.
   (vi) A hemigroup: k plays the part of e. $(k + 1) * x = k$ is not solvable for x: it would require $x = k - 1$. Therefore the hemigroup is not a group.

2. If u, v, and w are any three functions,

$$((u * v) * w)(x) = (u * v)(w(x)) = u(v(w(x))).$$

By similar reasoning, this equals $(u * (v * w))(x)$, and so $*$ is associative. $(f*f)(x) = 1/(1/x) = x$. Therefore $f*f$ plays the part of ε. It is easily verified that the product of any two of the six given elements is one or other of the six: and that f, g, $f*f$ and $g*f*g$ are their own inverses, and that $f*g$ and $g*f$ are inverses of one another.

3. If $a \mid b$ is any dyad, let $a = b*x$. Then $a \mid b = b*x \mid b*e = x \mid e$. Therefore the isomorphism $x \mid e \longleftrightarrow x$ is between the set of *all* dyads of H and the set H itself.

4. If $a_1 \ldots a_n$ are the elements of H, then $a_1 * a_i \ldots a_n * a_i$ are elements of H and, by **1c**, no two are the same. Thus they are the n elements of H, in some order. Thus if a_k is any element of H there is an a_j of H such that $a_j * a_i = a_k$—and this is true for each a_i. That is, the equation $x*a = b$ is always solvable. Similarly for $a*x = b$.

5. (i) $(y*x)*z = (z*x)*y$ by **2**.

(ii) The mapping is one-to-one, by the cancellation law.

$$(a*x \mid a)*(a*y \mid a) = a*(x*y) \mid a.$$

(iii) **2–10** do not require **1d**. The isomorphism in the restated **11** is (ii), which is, by (i), independent of a. Also $a \mid a$ is independent of a, and so can play the part of ε. **13** is clearly still true, and **14–16** follow from it. **17–22** follow from **16**.

6. $(\bar{\eta}*(\bar{\eta}*\eta))*\bar{\eta} = \varepsilon$. $(\bar{\eta}*\bar{\eta})*(\eta*\bar{\eta}) = \eta*\bar{\eta}$. This gives us **d**. Then $\eta*\varepsilon = \eta*\bar{\eta}\ *\eta = \varepsilon*\eta = \eta$. This is symmetric with **c**.

17 does not use **b**; and, by symmetry with it, $\xi*\alpha = \beta$ is solvable for ξ, and $\eta*\alpha = \xi*\alpha$ only if $\eta = \xi$.

18: if $\iota*\xi = \xi$ then $\iota = \varepsilon$ by **17**; and if $\alpha*\kappa = \varepsilon$ then $\kappa = \bar{\alpha}$ by **17**; and, by symmetry with these, if $\xi*\iota = \xi$ then $\iota = \varepsilon$, and if $\kappa*\alpha = \varepsilon$, then $\kappa = \bar{\alpha}$.

19 and **20** do not use **b**.

The theorem corresponding to **21** is $\overline{\xi*\eta} = \bar{\eta}*\bar{\xi}$.

C

1. $o' + o' = (0' - 0) + (0' - 0) = (0' \cdot 0' + 0 \cdot 0) - (0' \cdot 0 + 0 \cdot 0') = 0'' - 0$.

2. If $0'' - 0 = 0' - 0$, then $0'' + 0 = 0' + 0$, whence $0' + 0' = 0' + 0$, whence $0' = 0$, contradicting **A2**.

3. **A1a**$^\times$: if $\xi^\times = \eta^\times$, then $\xi + (-o') = \eta + (-o')$, and so $\xi = \eta$.

A1b$^\times$: o plays the part of 0, because if $\xi + (-o') = o$ then $-\xi$ is not positive.

A1c$^\times$: if N contains o and contains $\xi^\times$ whenever it contains ξ, then it contains $0 - 0$ and contains $0 - x'$ whenever it contains $0 - x$, whence it contains, by **A1c**, every $0 - x$.

A3a$^\times$ is $\xi + o = \xi$, and **A3b**$^\times$ is $\xi^\times + \eta = (\xi + \eta)^\times$. These are clearly satisfied and they *uniquely* define addition.

A9b$^\times$ implies $o^\times \cdot o^\times = o^\times$, which implies $(-o')\cdot(-o') = (-o')$, which is false, because $o' \neq -o'$, which can be proved as in exercise 2.

4. The correspondence is clearly one-to-one, and

$$(0 - x) + (0 - y) = 0 - (x + y).$$

$(0 - 0')\cdot(0 - 0') = (0' - 0) \neq (0 - 0')$ because $0' + 0' \neq 0$, as before.

5. If α is positive, put $\beta = o$: otherwise put $\beta = -\alpha$.

6. If $-\alpha$ is positive, put $\gamma = o$. If not, then put $\gamma = \alpha$ if β is positive, $\gamma = -\alpha$ if β is not positive; and use **C33**.

D

1. **1a, b, e, f,** and **k** are easily verified. **1c, d, g,** and **h** are true if $o = a$, $-a = a$, $-b = b$, $i = b$, and $b^- = b$.

2. $\{0, 1, 2\}$, with addition and multiplication modulo 3.

3. In the field of exercise 1, $i + i = o$.

Let ${}^m i$ denote the sum of m i's. Clearly ${}^p i \cdot {}^q i = {}^{p\cdot q} i$ and ${}^p i + {}^q i = {}^{p+q} i$. If $p\cdot q$ is the smallest m for which ${}^m i = o$, then ${}^p i \neq o$, and ${}^q i \neq o$; but ${}^p i \cdot {}^q i = {}^{p\cdot q} i = o$.

If ${}^m i \neq o$ for every non-zero m, then ${}^p i \longleftrightarrow p$ is a one-to-one correspondence, because if ${}^p i = {}^{p+q} i$ then ${}^q i = o$.

4. By **D1a, f, g,** and **k.**

$(i + i)\cdot(x + y) = (x + x) + (y + y) = (x + (x + y)) + y$, and
$(x + y)\cdot(i + i) = (x + y) + (x + y) = (x + (y + x)) + y$.
The result follows by adding $-y$ on the right and $-x$ on the left, because, by a proof similar to that in exercise **B**6, $z + o = z$ and $-z + z = o$.

E

1. If $\boldsymbol{x}$ is α/β, put $\boldsymbol{y} = \alpha/(\beta + (\beta + \beta))$.

2. If $(\alpha/\beta)^2 = 2$, where α/β is in its lowest terms, then $\alpha^2 = 2{\cdot}\beta^2$, and so α would be even, and so β would be odd. Then α^2 would be divisible by 4 but $2{\cdot}\beta^2$ would not.

3. If there is such an $\boldsymbol{x}$, let it be γ/δ where γ and δ have no common factor. Then $\gamma^2{\cdot}\beta = \delta^2{\cdot}\alpha$. α is prime to β and so divides γ^2: let $\gamma^2 = \xi{\cdot}\alpha$. Then $\delta^2 = \xi{\cdot}\beta$. Then $\xi = o'$; otherwise γ and δ would have a common factor. Put $\boldsymbol{y} = \gamma/o'$, and $\boldsymbol{z} = \delta/o'$.

   If there are such a $\boldsymbol{y}$ and $\boldsymbol{z}$, they are non-zero because α and β are. Put $\boldsymbol{x} = \boldsymbol{y}{\cdot}\boldsymbol{z}^-$.

4. Putting $o = (\boldsymbol{0}, \boldsymbol{0})$, $i = (\boldsymbol{1}, \boldsymbol{0})$, $-(\mathbf{a}, \boldsymbol{b}) = (-\mathbf{a}, -\boldsymbol{b})$, and $(\mathbf{a}, \boldsymbol{b})^- = (-\mathbf{a}\cdot\boldsymbol{c}, \boldsymbol{b}\cdot\boldsymbol{c})$, where $\boldsymbol{c} = (\boldsymbol{2}\cdot\boldsymbol{b}^2 - \mathbf{a}^2)^-$, whose existence is assured by exercise 2, **D1—a k** are easily verified.

   The given correspondence is clearly an isomorphism.

   $(\boldsymbol{0}, \boldsymbol{1}){\cdot}(\boldsymbol{0}, \boldsymbol{1}) = (\boldsymbol{2}, \boldsymbol{0})$.

F

1. $x + (-y) = -y + (-(-x))$.

2. If $y = o$, then $o \geqslant x$ and $x \geqslant o$, whence $x = o$. By exercise 1, $-y \geqslant -x$, whence $(-x)^- \geqslant (-y)^-$ by **14**, whence $-(x^-) \geqslant -(y^-)$.

3. If $i + a = o$, then $a = -i \notin P$.

4. **A1a**: if $x' = y'$, then $x + i = y + i$, and so $x = y$.

   A1b: $o \in P$, and if $x \in P$ then $x' \in P$; therefore every element of I is in P, whence, by exercise 3, if $a' = o$ then $a \notin P$.

   A1c: obvious.

   A3a and **b** and **A9a** and **b** are clearly satisfied by addition and multiplication in the field.

   A1b may fail for a non-ordered field: in exercise **D1**, $i' = o$.

5. Let $(a, b) \in P$ if and only if (i) $a \geqslant 0$ and $b \geqslant 0$,

   or (ii) $a \geqslant 0$ and $a^2 \geqslant 2b^2$,

   or (iii) $b \geqslant 0$ and $2b^2 \geqslant a^2$.

G

1. If $\boldsymbol{p}$ is ξ/η where η is positive, put $\alpha = \xi$ and $\beta = \eta$; if η is not positive, put $\alpha = -\xi$ and $\beta = -\eta$.

2. Let $\boldsymbol{p}$ be ξ/η and $\boldsymbol{q}$ be θ/ϕ where η and ϕ are positive (using exercise 1). Then put $\alpha = \xi \cdot \phi$, $\beta = \eta \cdot \theta$, and $\gamma = \eta \cdot \phi$.

3. Let $\boldsymbol{p}$ be α/γ and $\boldsymbol{q}$ be β/γ where γ is positive (using exercises 1 and 2). Then $(\alpha + (-\beta))/\gamma \geqslant 0$ and so is equal to σ/ρ for some positive σ and ρ. Then, by **E2** and **C32**, $\alpha + (-\beta)$ is positive. Now put $\boldsymbol{r} = (\alpha + \beta)/(\gamma + \gamma)$.

4. If $\boldsymbol{x}$ is positive, put $\boldsymbol{y} = \boldsymbol{x}$ and $\boldsymbol{z} = \boldsymbol{0}$; if not, put $\boldsymbol{y} = \boldsymbol{0}$ and $\boldsymbol{z} = -\boldsymbol{x}$.

5. By various isomorphisms, the positive integers form a succession. Let $\boldsymbol{N}$ be the set of all positive integers $\boldsymbol{x}$ such that if $\boldsymbol{y} \in \boldsymbol{M}$ then $\boldsymbol{y} \geqslant \boldsymbol{x}$. Then $\boldsymbol{0} \in \boldsymbol{N}$. If $\boldsymbol{x}' \in \boldsymbol{N}$ whenever $\boldsymbol{x} \in \boldsymbol{N}$ then every positive integer would be in $\boldsymbol{N}$, which cannot be. Thus there is an $\boldsymbol{x}$ such that $\boldsymbol{x} \in \boldsymbol{N}$ and $\boldsymbol{x}' \notin \boldsymbol{N}$. This $\boldsymbol{x}$ is easily seen to be the required $\boldsymbol{k}$.

6. If $\boldsymbol{x}$ is positive, let $\boldsymbol{M}$ be the set of all positive integers $\boldsymbol{m}$ such that $\boldsymbol{m} \geqslant \boldsymbol{x}$. Let $\boldsymbol{k}$ be as in exercise 5, and then let $\boldsymbol{p}$ be such that $\boldsymbol{p} + \boldsymbol{1} = \boldsymbol{k}$. Now if also $\boldsymbol{q} + \boldsymbol{1} \geqslant \boldsymbol{x} \geqslant \boldsymbol{q}$, then $\boldsymbol{q} + \boldsymbol{1} \in \boldsymbol{M}$, and so $\boldsymbol{q} + \boldsymbol{1} \geqslant \boldsymbol{p} + \boldsymbol{1}$. If $\boldsymbol{q} \neq \boldsymbol{p}$, then $\boldsymbol{q} \geqslant \boldsymbol{p} + \boldsymbol{1}$ and so $\boldsymbol{q} \geqslant \boldsymbol{x}$, which is not so.

   If $\boldsymbol{x}$ is not positive there is, by the above result, one and only one integral rational $\boldsymbol{r}$ such that $\boldsymbol{r} + \boldsymbol{1} \geqslant -\boldsymbol{x} \geqslant \boldsymbol{r}$. Then $\boldsymbol{p}$ is the number for which $\boldsymbol{p} + \boldsymbol{1} = -\boldsymbol{r}$.

7. If there is such an $\boldsymbol{r}$, then $\boldsymbol{q} + (-\boldsymbol{p}) = \boldsymbol{r} \in P$.

   If $\boldsymbol{q} \geqslant \boldsymbol{p}$, let $\boldsymbol{r}$ be $\boldsymbol{q} + (-\boldsymbol{p})$, which is in P.

H

1. If $a \cdot x^2 + 2h \cdot x + b = 0$, put $y = a \cdot x + h$.

   If $y^2 = h^2 - a \cdot b$, put $x = a^{-} \cdot (y - h)$.

2. $x^2 \geqslant 0$. Therefore $x^{2p} \geqslant 0$. $-x \geqslant 0$. Therefore $-x^{2p+1} \geqslant 0$.

3. See **J22**.

4. If $x > 1$ then, by exercise 3, $x^p \geqslant 1 + p\cdot(x-1)$. Choose p so that $p\cdot(x-1) \geqslant k$.

If $1 > x > 0$, there is a q for which $(x^-)^q \geqslant k$. Put $p = -q$.

If $0 > x$, $(-x)^r \geqslant k$. If r is even, $x^r = (-x)^r$. If not, either $x^{r+1} \geqslant (-x)^r$ or $x^{r-1} \geqslant (-x)^r$.

5. Let t^{n+1} be the smallest power of t greater than a. Dividing a by t^n we have $a = x_n\cdot t^n + r_n$ where $t > x_n > 0$ and $t^n > r_n \geqslant 0$. Now divide r_n by t^{n-1}, getting $r_n = x_{n-1}\cdot t^{n-1} + r_{n-1}$, and so on. Now if $a = \sum_{p=0}^{m} y_p\cdot t^p$, clearly t^{m+1} is the smallest power of t greater than a, and so $m = n$. Then $x_n\cdot t^n + r_n = y_n\cdot t^n + s_n$ where $t^n > s_n \geqslant 0$. Then $(x_n - y_n)\cdot t^n = s_n - r_n$. But $x_n - y_n$ is an integer, and so can only be 0. And so on.

6. If **a** holds, $0^0 = 1$. If **b** holds, $0^0 = 0^{-1+1} = 0^{-1}\cdot 0 = 0$.

I

1. Let k be as in lemma **15**, and l similarly for $\{u\}$. If $b > 0$, then $u_p - u_q \leqslant \frac{1}{2}b\cdot k^{-1}$ and $x_p - x_q \leqslant \frac{1}{2}b\cdot l^{-1}$ for all large p and q. And $|\, u_p\cdot x_p - u_q\cdot x_q \,| \leqslant |\, x_p \,|\cdot|\, u_p - u_q \,| + |\, u_q \,|\cdot|\, x_p - x_q \,|$.
Thus $\{u\cdot x\}$ is a Cauchy sequence.

$|\, u_p\cdot x_p - v_p\cdot x_p \,| \leqslant |\, x_p \,|\cdot|\, u_p - v_p \,| \leqslant b$ for all large p. Thus $\{u\cdot x\}$ and $\{v\cdot x\}$ are in the same Cauchy number.

Applying this result again we see that $\{v\cdot x\}$ and $\{v\cdot y\}$ are in the same Cauchy number.

2. Let k be such that $\frac{1}{2}k\cdot b^{-1} \geqslant y_p$ for every p (**I15**). Then
$k \geqslant \frac{1}{2}k\cdot b^{-1}\cdot b + \frac{1}{2}k\cdot b^{-1}\cdot b \geqslant |\, x_p\cdot y_p \,| + |\, x_q\cdot y_q \,| \geqslant |\, x_p\cdot y_p - x_q\cdot y_q \,|$
for all large p and q.

If $\{y\}$ has property (α) then $|\, x_p + y_p - (x_q + y_q) \,| \leqslant b$.

If $\{x + y\}$ has property (α) then $|\, (x_p + y_p) - (x_q + y_q) \,| \leqslant \frac{1}{2}b$, and so $|\, y_p - y_q \,| \leqslant \frac{1}{2}b + |\, x_p \,| + |\, x_q \,|$.

3. It is easy to prove that, given l, $n_p \geqslant l$ for all large p. Hence if $\boldsymbol{P}(p)$ is true for all large p, so is $\boldsymbol{P}(n_p)$.

If $d > 0$, then $d \geqslant |\, a_{n_p} - a_{n_q} \,|$ for all large p and q. Therefore $\{b\}$ is a Cauchy sequence. $d \geqslant |\, a_{n_p} - a_p \,|$ for all large p. Therefore $\{a\}$ and $\{b\}$ are in the same Cauchy number.

4. Let $\mathbf{x}$ be a Cauchy number and $\{x\}$ any Cauchy sequence in $\mathbf{x}$. Let $k \geqslant x_p$ for every p. Then $\tilde{\mathbf{k}} \geqslant \mathbf{x}$. If $n \geqslant k$, then $\tilde{\mathbf{n}} \geqslant \mathbf{x}$.

J

1. It is easy to verify that F is an ordered field.

$r/1 \geqslant x/1$ if and only if $(-x + r)/1$ is positive, i.e. $-1 \geqslant 0$, which is not so.

2. Let p be an integer and an upper bound of M. Let $m \in M$ and let q be an integer such that $m \geqslant -q$.

For any whole number r, as s ranges from 0 to $(p + q)\cdot 2^r$, $-q + 2^{-r}\cdot s$ ranges from $-q$ to p and so some of its values are upper bounds of M. Let l_r be the smallest. Then $l_r - 2^{-r}$ is not an upper bound and so, whenever $t \geqslant r$, $l_r \geqslant l_t \geqslant l_r - 2^{-r}$. It soon follows that $\{l\}$ is a Cauchy sequence and that if l is its limit, then $l_r \geqslant l \geqslant l_r - 2^{-r}$ for every r.

l is an upper bound, for if $m > l$ then $m - l > 2^r$ for some r, whence $m > l_r$ because $l \geqslant l_r - 2^{-r}$.

Now let k be any upper bound. If $l > k$, let $l - k > 2^{-r}$. $l^r - 2^{-r}$ is not an upper bound and so $m > l_r - 2^{-r}$ for some m. Then $k > l_r - 2^{-r}$ and so $l > l_r$, which is not so.

3. Let l be the least upper bound of Q. If $x > l$, then $x \notin Q$. Therefore $x \in P$. If $l > x$, then x is not an upper bound of Q. Therefore there is a y in Q such that $y > x$. But if $x \in P$, then $x \geqslant y$. Therefore $x \in Q$.

Let $\boldsymbol{P}$ be the set of all rational numbers $\boldsymbol{x}$ such that $\boldsymbol{x}^2 \geqslant \boldsymbol{2}$, and $\boldsymbol{Q}$ be the set of all rational numbers $\boldsymbol{x}$ such that $\boldsymbol{2} \geqslant \boldsymbol{x}^2$. By the proof of **H9**, $\boldsymbol{P}$ and $\boldsymbol{Q}$ are seen to be non-null. By exercise **E2**, no rational number is in both; and clearly each rational number is in one or the other.

Now if a rational number $\boldsymbol{k}$ with the given properties exists and $\boldsymbol{k}^2 > \boldsymbol{2}$, we can find a $\boldsymbol{p}$ in $\boldsymbol{P}$ such that $|\boldsymbol{p}^2 - \boldsymbol{2}| \leqslant |\boldsymbol{k}^2 - \boldsymbol{2}|$ by letting $\boldsymbol{b} = \frac{1}{4}|\boldsymbol{k}^2 - \boldsymbol{2}|$ and $\boldsymbol{p}$ be the first of the numbers $\boldsymbol{1}, \boldsymbol{1} + \boldsymbol{b}, \boldsymbol{1} + \boldsymbol{2b}, \ldots$ in $\boldsymbol{P}$. But then $\boldsymbol{k} > \boldsymbol{p}$, which is not so. Similarly we cannot have $\boldsymbol{2} > \boldsymbol{k}^2$. Since $\boldsymbol{k}^2 \neq \boldsymbol{2}$, $\boldsymbol{k}$ cannot exist.

4. $|\mathbf{1} - \cdot 99 \ldots 9| = 10^{-n}$, which is as small as we please for all large n. Therefore $\cdot\dot{9}$ represents 1. Then $a_0\cdot a_1 a_2 \ldots a_n\dot{9} = a_0\cdot a_1 a_2 \ldots b_n\dot{0}$ if $b_n = a_n + 1$.

If $\cdot a_1 a_2 a_3 \ldots = x = \cdot b_1 b_2 b_3 \ldots$ and $a_1 > b_1$, then $b_1 + 1 \geqslant 10x \geqslant a_1$. But a_1 and b_1 are integers, and so $a_1 = b_1 + 1$. Then $1\cdot a_2 a_3 \ldots = \cdot b_2 b_3 \ldots.$ Then $a_2 = a_3 = \ldots = 0$, for if not $1\cdot a_2 a_3 \ldots > 1 = \cdot\dot{9} \geqslant \cdot b_2 b_3 \ldots.$ And $b_2 = b_3 = \ldots = 9$, for if not $1\cdot a_2 a_3 \ldots \geqslant 1 = \dot{9} > \cdot b_2 b_3 \ldots.$

It follows easily that if $a_0\cdot a_2 a_3 \ldots = b_0\cdot b_1 b_2 \ldots$ and a_j and b_j are the first unequal corresponding digits, then *either* $a_j = b_j + 1$, $a_{j+1} \ldots = \dot{0}$, and $b_{j+1} \ldots = \dot{9}$, *or* $b_j = a_j + 1$, etc.

INDEX

CATALOGUE OF DOVER BOOKS

PHILOSOPHY OF SCIENCE AND MATHEMATICS

FOUNDATIONS OF SCIENCE: THE PHILOSOPHY OF THEORY AND EXPERIMENT, N. R. Campbell. A critique of the most fundamental concepts of science in general and physics in particular. Examines why certain propositions are accepted without question, demarcates science from philosophy, clarifies the understanding of the tools of science. Part One analyzes the presuppositions of scientific thought: existence of the material world, nature of scientific laws, multiplication of probabilities, etc.: Part Two covers the nature of experiment and the application of mathematics: conditions for measurement, relations between numerical laws and theories, laws of error, etc. An appendix covers problems arising from relativity, force, motion, space, and time. A classic in its field. Index. xiii + 565pp. 5⅝ x 8⅜.
S372 Paperbound **$2.95**

THE NATURE OF PHYSICAL THEORY, P. W. Bridgman. Here is how modern physics looks to a highly unorthodox physicist—a Nobel laureate. Pointing out many absurdities of science, and demonstrating the inadequacies of various physical theories, Dr. Bridgman weighs and analyzes the contributions of Einstein, Bohr, Newton, Heisenberg, and many others. This is a non-technical consideration of the correlation of science and reality. Index. xi + 138pp. 5⅜ x 8.
S33 Paperbound **$1.25**

THE VALUE OF SCIENCE, Henri Poincaré. Many of the most mature ideas of the "last scientific universalist" covered with charm and vigor for both the beginning student and the advanced worker. Discusses the nature of scientific truth, whether order is innate in the universe or imposed upon it by man, logical thought versus intuition (relating to math, through the works of Weierstrass, Lie, Klein, Riemann), time and space (relativity, psychological time, simultaneity), Hertz's concept of force, interrelationship of mathematical physics to pure math, values within disciplines of Maxwell, Carnot, Mayer, Newton, Lorentz, etc. Index. iii + 147pp. 5⅜ x 8.
S469 Paperbound **$1.35**

SCIENCE AND HYPOTHESIS, Henri Poincaré. Creative psychology in science. How such concepts as number, magnitude, space, force, classical mechanics were developed, and how the modern scientist uses them in his thought. Hypothesis in physics, theories of modern physics. Introduction by Sir James Larmor. "Few mathematicians have had the breadth of vision of Poincaré, and none is his superior in the gift of clear exposition," E. T. Bell. Index. 272pp. 5⅜ x 8.
S221 Paperbound **$1.35**

PHILOSOPHY AND THE PHYSICISTS, L. S. Stebbing. The philosophical aspects of modern science examined in terms of a lively critical attack on the ideas of Jeans and Eddington. Discusses the task of science, causality, determinism, probability, consciousness, the relation of the world of physics to that of everyday experience. Probes the philosophical significance of the Planck-Bohr concept of discontinuous energy levels, the inferences to be drawn from Heisenberg's Uncertainty Principle, the implications of "becoming" involved in the 2nd law of thermodynamics, and other problems posed by the discarding of Laplacean determinism. 285pp. 5⅜ x 8.
T480 Paperbound **$1.65**

THE PHILOSOPHICAL WRITINGS OF PEIRCE, edited by Justus Buchler. (Formerly published as THE PHILOSOPHY OF PEIRCE.) This is a carefully balanced exposition of Peirce's complete system, written by Peirce himself. It covers such matters as scientific method, pure chance vs. law, symbolic logic, theory of signs, pragmatism, experiment, and other topics. Introduction by Justus Buchler, Columbia University. xvi + 368pp. 5⅜ x 8.
T217 Paperbound **$2.00**

LANGUAGE, TRUTH AND LOGIC, A. Ayer. A clear introduction to the Vienna and Cambridge schools of Logical Positivism. It sets up specific tests by which you can evaluate validity of ideas, etc. Contents: Function of philosophy, elimination of metaphysics, nature of analysis, a priori, truth and probability, etc. 10th printing. "I should like to have written it myself," Bertrand Russell. Index. 160pp. 5⅜ x 8.
T10 Paperbound **$1.25**

MATHEMATICS AND SCIENCE: LAST ESSAYS (DERNIÈRES PENSÉES), Henri Poincaré. Translated by J. W. Bolduc. A posthumous volume of articles and lectures by the great French mathematician, philosopher, scientist. Here are nine pieces, never before translated into English, on such subjects as The Evolution of Laws, Space and Time, Space and 3 Dimensions, The Logic of infinity in Mathematics (discussing Russell's theory of types), Mathematics and Logic, The Quantum Theory and its Modern Applications, Relationship Between Matter and Ether, Ethics and Science and The Moral Alliance. First English translation of Dernières Pensées. New index. viii + 128pp. 5⅜ x 8½.
S1101 Paperbound **$1.25**

THE PSYCHOLOGY OF INVENTION IN THE MATHEMATICAL FIELD, J. Hadamard. Where do ideas come from? What role does the unconscious play? Are ideas best developed by mathematical reasoning, word reasoning, visualization? What are the methods used by Einstein, Poincaré, Galton, Riemann? How can these techniques be applied by others? Hadamard, one of the world's leading mathematicians, discusses these and other questions. xiii + 145pp. 5⅜ x 8.
T107 Paperbound **$1.25**

MATHEMATICS, ELEMENTARY TO INTERMEDIATE

HOW TO CALCULATE QUICKLY, Henry Sticker. This handy volume offers a tried and true method for helping you in the basic mathematics of daily life—addition, subtraction, multiplication, division, fractions, etc. It is designed to awaken your "number sense" or the ability to see relationships between numbers as whole quantities. It is not a collection of tricks working only on special numbers, but a serious course of over 9,000 problems and their solutions, teaching special techniques not taught in schools: left-to-right multiplication, new fast ways of division, etc. 5 or 10 minutes daily use will double or triple your calculation speed. Excellent for the scientific worker who is at home in higher math, but is not satisfied with his speed and accuracy in lower mathematics. 256pp. 5 x 7¼. T295 Paperbound **$1.00**

TEACH YOURSELF books. For adult self-study, for refresher and supplementary study.

The most effective series of home study mathematics books on the market! With absolutely no outside help, they will teach you as much as any similar college or high-school course, or will helpfully supplement any such course. Each step leads directly to the next, each question is anticipated. Numerous lucid examples and carefully-wrought practice problems illustrate meanings. Not skimpy outlines, not surveys, not usual classroom texts, these 204- to 380-page books are packed with the finest instruction you'll find anywhere for adult self-study.

TEACH YOURSELF ALGEBRA, P. Abbott. Formulas, coordinates, factors, graphs of quadratic functions, quadratic equations, logarithms, ratio, irrational numbers, arithmetical, geometrical series, much more. 1241 problems, solutions. Tables. 52 illus. 307pp. 6⅞ x 4¼. Clothbound **$2.00**

TEACH YOURSELF GEOMETRY, P. Abbott. Solids, lines, points, surfaces, angle measurement, triangles, theorem of Pythagoras, polygons, loci, the circle, tangents, symmetry, solid geometry, prisms, pyramids, solids of revolution, etc. 343 problems, solutions. 268 illus. 334pp. 6⅞ x 4¼. Clothbound **$2.00**

TEACH YOURSELF TRIGONOMETRY, P. Abbott. Geometrical foundations, indices, logarithms, trigonometrical ratios, relations between sides, angles of triangle, circular measure, trig. ratios of angles of any magnitude, much more. Requires elementary algebra, geometry. 465 problems, solutions. Tables. 102 illus. 204pp. 6⅞ x 4¼. Clothbound **$2.00**

TEACH YOURSELF THE CALCULUS, P. Abbott. Variations in functions, differentiation, solids of revolution, series, elementary differential equations, areas by integral calculus, much more. Requires algebra, trigonometry. 970 problems, solutions. Tables. 89 illus. 380pp. 6⅞ x 4¼. Clothbound **$2.00**

TEACH YOURSELF THE SLIDE RULE, B. Snodgrass. Fractions, decimals, A-D scales, log-log scales, trigonometrical scales, indices, logarithms. Commercial, precision, electrical, dualistic, Brighton rules. 80 problems, solutions. 10 illus. 207pp. 6⅞ x 4¼. Clothbound **$2.00**

ARITHMETICAL EXCURSIONS: AN ENRICHMENT OF ELEMENTARY MATHEMATICS, H. Bowers and J. Bowers. For students who want unusual methods of arithmetic never taught in school; for adults who want to increase their number sense. Little known facts about the most simple numbers, arithmetical entertainments and puzzles, figurate numbers, number chains, mysteries and folklore of numbers, the "Hin-dog-abic" number system, etc. First publication. Index. 529 numbered problems and diversions, all with answers. Bibliography. 50 figures. xiv + 320pp. 5⅜ x 8. T770 Paperbound **$1.65**

HOW DO YOU USE A SLIDE RULE? by A. A. Merrill. Not a manual for mathematicians and engineers, but a lucid step-by-step explanation that presents the fundamental rules clearly enough to be understood by anyone who could benefit by the use of a slide rule in his work or business. This work concentrates on the 2 most important operations: multiplication and division. 10 easy lessons, each with a clear drawing, will save you countless hours in your banking, business, statistical, and other work. First publication. Index. 2 Appendixes. 10 illustrations. 78 problems, all with answers. vi + 36pp. 6⅛ x 9¼. T62 Paperbound **60¢**

THE THEORY AND OPERATION OF THE SLIDE RULE, J. P. Ellis. Not a skimpy "instruction manual", but an exhaustive treatment that will save you hours throughout your career. Supplies full understanding of every scale on the Log Log Duplex Decitrig type of slide rule. Shows the most time-saving methods, and provides practice useful in the widest variety of actual engineering situations. Each operation introduced in terms of underlying logarithmic theory. Summary of prerequisite math. First publication. Index. 198 figures. Over 450 problems with answers. Bibliography. 12 Appendices. ix + 289pp. 5⅜ x 8. S727 Paperbound **$1.50**

COLLEGE ALGEBRA, H. B. Fine. Standard college text that gives a systematic and deductive structure to algebra; comprehensive, connected, with emphasis on theory. Discusses the commutative, associative, and distributive laws of number in unusual detail, and goes on with undetermined coefficients, quadratic equations, progressions, logarithms, permutations, probability, power series, and much more. Still most valuable elementary-intermediate text on the science and structure of algebra. Index. 1560 problems, all with answers. x + 631pp. 5⅜ x 8. T211 Paperbound **$2.50**

COORDINATE GEOMETRY, L. P. Eisenhart. Thorough, unified introduction. Unusual for advancing in dimension within each topic (treats together circle, sphere; polar coordinates, 3-dimensional coordinate systems; conic sections, quadric surfaces), affording exceptional insight into subject. Extensive use made of determinants, though no previous knowledge of them is assumed. Algebraic equations of 1st degree, 2 and 3 unknowns, carried further than usual in algebra courses. Over 500 exercises. Introduction. Appendix. Index. Bibliography. 43 illustrations. 310pp. 5⅜ x 8. S600 Paperbound **$1.65**

A TREATISE ON PLANE AND ADVANCED TRIGONOMETRY, E. W. Hobson. Extraordinarily wide coverage, going beyond usual college level trig, one of the few works covering advanced trig in full detail. By a great expositor with unerring anticipation and lucid clarification of potentially difficult points. Includes circular functions; expansion of functions of multiple angle; trig tables; relations between sides and angles of triangle; complex numbers; etc. Many problems solved completely. "The best work on the subject." Nature. Formerly entitled "A Treatise on Plane Trigonometry." 689 examples. 6 figures. xvi + 383pp. 5⅜ x 8.
S353 Paperbound **$2.25**

FAMOUS PROBLEMS OF ELEMENTARY GEOMETRY, Felix Klein. Expanded version of the 1894 Easter lectures at Göttingen. 3 problems of classical geometry, in an excellent mathematical treatment by a famous mathematician: squaring the circle, trisecting angle, doubling cube. Considered with full modern implications: transcendental numbers, pi, etc. Notes by R. Archibald. 16 figures. xi + 92pp. 5⅜ x 8. T298 Paperbound **$1.00**

MONOGRAPHS ON TOPICS OF MODERN MATHEMATICS, edited by **J. W. A. Young.** Advanced mathematics for persons who haven't gone beyond or have forgotten high school algebra. 9 monographs on foundation of geometry, modern pure geometry, non-Euclidean geometry, fundamental propositions of algebra, algebraic equations, functions, calculus, theory of numbers, etc. Each monograph gives proofs of important results, and descriptions of leading methods, to provide wide coverage. New introduction by Prof. M. Kline, N. Y. University. 100 diagrams. xvi + 416pp. 6⅛ x 9¼. S289 Paperbound **$2.00**

HIGHER MATHEMATICS FOR STUDENTS OF CHEMISTRY AND PHYSICS, J. W. Mellor. Not abstract, but practical, building its problems out of familiar laboratory material, this covers differential calculus, coordinate, analytical geometry, functions, integral calculus, infinite series, numerical equations, differential equations, Fourier's theorem, probability, theory of errors, calculus of variations, determinants. "If the reader is not familiar with this book, it will repay him to examine it," CHEM. & ENGINEERING NEWS. 800 problems. 189 figures. Bibliography. xxi + 641pp. 5⅜ x 8. S193 Paperbound **$2.50**

TRIGONOMETRY REFRESHER FOR TECHNICAL MEN, A. Albert Klaf. 913 detailed questions and answers cover the most important aspects of plane and spherical trigonometry. They will help you to brush up or to clear up difficulties in special areas. The first portion of this book covers plane trigonometry, including angles, quadrants, trigonometrical functions, graphical representation, interpolation, equations, logarithms, solution of triangle, use of the slide rule and similar topics. 188 pages then discuss application of plane trigonometry to special problems in navigation, surveying, elasticity, architecture, and various fields of engineering. Small angles, periodic functions, vectors, polar coordinates, de Moivre's theorem are fully examined. The third section of the book then discusses spherical trigonometry and the solution of spherical triangles, with their applications to terrestrial and astronomical problems. Methods of saving time with numerical calculations, simplification of principal functions of angle, much practical information make this a most useful book. 913 questions answered. 1738 problems, answers to odd numbers. 494 figures. 24 pages of useful formulae, functions. Index. x + 629pp. 5⅜ x 8. T371 Paperbound **$2.00**

TEXTBOOK OF ALGEBRA, G. Chrystal. One of the great mathematical textbooks, still about the best source for complete treatments of the topics of elementary algebra; a chief reference work for teachers and students of algebra in advanced high school and university courses, or for the mathematician working on problems of elementary algebra or looking for a background to more advanced topics. Ranges from basic laws and processes to extensive examination of such topics as limits, infinite series, general properties of integral numbers, and probability theory. Emphasis is on algebraic form, the foundation of analytical geometry and the key to modern developments in algebra. Prior course in algebra is desirable, but not absolutely necessary. Includes theory of quotients, distribution of products, arithmetical theory of surds, theory of interest, permutations and combinations, general expansion theorems, recurring fractions, and much, much more. Two volume set. Index in each volume. Over 1500 exercises, approximately half with answers. Total of xlviii + 1187pp. 5⅜ x 8.
S750 Vol I Paperbound **$2.35**
S751 Vol II Paperbound **$2.35**
The set **$4.70**

MATHEMATICS—INTERMEDIATE TO ADVANCED

General

INTRODUCTION TO APPLIED MATHEMATICS, Francis D. Murnaghan. A practical and thoroughly sound introduction to a number of advanced branches of higher mathematics. Among the selected topics covered in detail are: vector and matrix analysis, partial and differential equations, integral equations, calculus of variations, Laplace transform theory, the vector triple product, linear vector functions, quadratic and bilinear forms, Fourier series, spherical harmonics, Bessel functions, the Heaviside expansion formula, and many others. Extremely useful book for graduate students in physics, engineering, chemistry, and mathematics. Index. 111 study exercises with answers. 41 illustrations. ix + 389pp. 5⅜ x 8½.
S1042 Paperbound **$2.25**

OPERATIONAL METHODS IN APPLIED MATHEMATICS, H. S. Carslaw and J. C. Jaeger. Explanation of the application of the Laplace Transformation to differential equations, a simple and effective substitute for more difficult and obscure operational methods. Of great practical value to engineers and to all workers in applied mathematics. Chapters on: Ordinary Linear Differential Equations with Constant Coefficients;; Electric Circuit Theory; Dynamical Applications; The Inversion Theorem for the Laplace Transformation; Conduction of Heat; Vibrations of Continuous Mechanical Systems; Hydrodynamics; Impulsive Functions; Chains of Differential Equations; and other related matters. 3 appendices. 153 problems, many with answers. 22 figures. xvi + 359pp. 5⅜ x 8½. S1011 Paperbound **$2.25**

APPLIED MATHEMATICS FOR RADIO AND COMMUNICATIONS ENGINEERS, C. E. Smith. No extraneous material here!—only the theories, equations, and operations essential and immediately useful for radio work. Can be used as refresher, as handbook of applications and tables, or as full home-study course. Ranges from simplest arithmetic through calculus, series, and wave forms, hyperbolic trigonometry, simultaneous equations in mesh circuits, etc. Supplies applications right along with each math topic discussed. 22 useful tables of functions, formulas, logs, etc. Index. 166 exercises, 140 examples, all with answers. 95 diagrams. Bibliography. x + 336pp. 5⅜ x 8. S141 Paperbound **$1.75**

Algebra, group theory, determinants, sets, matrix theory

ALGEBRAS AND THEIR ARITHMETICS, L. E. Dickson. Provides the foundation and background necessary to any advanced undergraduate or graduate student studying abstract algebra. Begins with elementary introduction to linear transformations, matrices, field of complex numbers; proceeds to order, basal units, modulus, quaternions, etc.; develops calculus of linears sets, describes various examples of algebras including invariant, difference, nilpotent, semi-simple. "Makes the reader marvel at his genius for clear and profound analysis," Amer. Mathematical Monthly. Index. xii + 241pp. 5⅜ x 8. S616 Paperbound **$1.50**

THE THEORY OF EQUATIONS WITH AN INTRODUCTION TO THE THEORY OF BINARY ALGEBRAIC FORMS, W. S. Burnside and A. W. Panton. Extremely thorough and concrete discussion of the theory of equations, with extensive detailed treatment of many topics curtailed in later texts. Covers theory of algebraic equations, properties of polynomials, symmetric functions, derived functions, Horner's process, complex numbers and the complex variable, determinants and methods of elimination, invariant theory (nearly 100 pages), transformations, introduction to Galois theory, Abelian equations, and much more. Invaluable supplementary work for modern students and teachers. 759 examples and exercises. Index in each volume. Two volume set. Total of xxiv + 604pp. 5⅜ x 8. S714 Vol I Paperbound **$1.85**
S715 Vol II Paperbound **$1.85**
The set **$3.70**

COMPUTATIONAL METHODS OF LINEAR ALGEBRA, V. N. Faddeeva, translated by **C. D. Benster.** First English translation of a unique and valuable work, the only work in English presenting a systematic exposition of the most important methods of linear algebra—classical and contemporary. Shows in detail how to derive numerical solutions of problems in mathematical physics which are frequently connected with those of linear algebra. Theory as well as individual practice. Part I surveys the mathematical background that is indispensable to what follows. Parts II and III, the conclusion, set forth the most important methods of solution, for both exact and iterative groups. One of the most outstanding and valuable features of this work is the 23 tables, double and triple checked for accuracy. These tables will not be found elsewhere. Author's preface. Translator's note. New bibliography and index. x + 252pp. 5⅜ x 8. S424 Paperbound **$2.00**

ALGEBRAIC EQUATIONS, E. Dehn. Careful and complete presentation of Galois' theory of algebraic equations; theories of Lagrange and Galois developed in logical rather than historical form, with a more thorough exposition than in most modern books. Many concrete applications and fully-worked-out examples. Discusses basic theory (very clear exposition of the symmetric group); isomorphic, transitive, and Abelian groups; applications of Lagrange's and Galois' theories; and much more. Newly revised by the author. Index. List of Theorems. xi + 208pp. 5⅜ x 8. S697 Paperbound **$1.45**

ALGEBRAIC THEORIES, L. E. Dickson. Best thorough introduction to classical topics in higher algebra develops theories centering around matrices, invariants, groups. Higher algebra, Galois theory, finite linear groups, Klein's icosahedron, algebraic invariants, linear transformations, elementary divisors, invariant factors; quadratic, bi-linear, Hermitian forms, singly and in pairs. Proofs rigorous, detailed; topics developed lucidly, in close connection with their most frequent mathematical applications. Formerly "Modern Algebraic Theories." 155 problems. Bibliography. 2 indexes. 285pp. 5⅜ x 8. S547 Paperbound **$1.50**

LECTURES ON THE ICOSAHEDRON AND THE SOLUTION OF EQUATIONS OF THE FIFTH DEGREE, Felix Klein. The solution of quintics in terms of rotation of a regular icosahedron around its axes of symmetry. A classic & indispensable source for those interested in higher algebra, geometry, crystallography. Considerable explanatory material included. 230 footnotes, mostly bibliographic. 2nd edition, xvi + 289pp. 5⅜ x 8. S314 Paperbound **$2.25**

LINEAR GROUPS, WITH AN EXPOSITION OF THE GALOIS FIELD THEORY, L. E. Dickson. The classic exposition of the theory of groups, well within the range of the graduate student. Part I contains the most extensive and thorough presentation of the theory of Galois Fields available, with a wealth of examples and theorems. Part II is a full discussion of linear groups of finite order. Much material in this work is based on Dickson's own contributions. Also includes expositions of Jordan, Lie, Abel, Betti-Mathieu, Hermite, etc. "A milestone in the development of modern algebra," W. Magnus, in his historical introduction to this edition. Index. xv + 312pp. 5⅜ x 8. S482 Paperbound **$1.95**

INTRODUCTION TO THE THEORY OF GROUPS OF FINITE ORDER, R. Carmichael. Examines fundamental theorems and their application. Beginning with sets, systems, permutations, etc., it progresses in easy stages through important types of groups: Abelian, prime power, permutation, etc. Except 1 chapter where matrices are desirable, no higher math needed. 783 exercises, problems. Index. xvi + 447pp. 5⅜ x 8. S300 Paperbound **$2.25**

THEORY OF GROUPS OF FINITE ORDER, W. Burnside. First published some 40 years ago, this is still one of the clearest introductory texts. Partial contents: permutations, groups independent of representation, composition series of a group, isomorphism of a group with itself, Abelian groups, prime power groups, permutation groups, invariants of groups of linear substitution, graphical representation, etc. 45pp. of notes. Indexes. xxiv + 512pp. 5⅜ x 8. S38 Paperbound **$2.75**

CONTINUOUS GROUPS OF TRANSFORMATIONS, L. P. Eisenhart. Intensive study of the theory and geometrical applications of continuous groups of transformations; a standard work on the subject, called forth by the revolution in physics in the 1920's. Covers tensor analysis, Riemannian geometry, canonical parameters, transitivity, imprimitivity, differential invariants, the algebra of constants of structure, differential geometry, contact transformations, etc. "Likely to remain one of the standard works on the subject for many years . . . principal theorems are proved clearly and concisely, and the arrangement of the whole is coherent," MATHEMATICAL GAZETTE. Index. 72-item bibliography. 185 exercises. ix + 301pp. 5⅜ x 8. S781 Paperbound **$2.00**

THE THEORY OF GROUPS AND QUANTUM MECHANICS, H. Weyl. Discussions of Schroedinger's wave equation, de Broglie's waves of a particle, Jordan-Hoelder theorem, Lie's continuous groups of transformations, Pauli exclusion principle, quantization of Maxwell-Dirac field equations, etc. Unitary geometry, quantum theory, groups, application of groups to quantum mechanics, symmetry permutation group, algebra of symmetric transformation, etc. 2nd revised edition. Bibliography. Index. xxii + 422pp. 5⅜ x 8. S269 Paperbound **$2.35**

APPLIED GROUP-THEORETIC AND MATRIX METHODS, Bryan Higman. The first systematic treatment of group and matrix theory for the physical scientist. Contains a comprehensive, easily-followed exposition of the basic ideas of group theory (realized through matrices) and its applications in the various areas of physics and chemistry: tensor analysis, relativity, quantum theory, molecular structure and spectra, and Eddington's quantum relativity. Includes rigorous proofs available only in works of a far more advanced character. 34 figures, numerous tables. Bibliography. Index. xiii + 454pp. 5⅜ x 8⅜. S1147 Paperbound **$3.00**

THE THEORY OF GROUP REPRESENTATIONS, Francis D. Murnaghan. A comprehensive introduction to the theory of group representations. Particular attention is devoted to those groups—mainly the symmetric and rotation groups—which have proved to be of fundamental significance for quantum mechanics (esp. nuclear physics). Also a valuable contribution to the literature on matrices, since the usual representations of groups are groups of matrices. Covers the theory of group integration (as developed by Schur and Weyl), the theory of 2-valued or spin representations, the representations of the symmetric group, the crystallographic groups, the Lorentz group, reducibility (Schur's lemma, Burnside's Theorem, etc.), the alternating group, linear groups, the orthogonal group, etc. Index. List of references. xi + 369pp. 5⅜ x 8½. S1112 Paperbound **$2.35**

THEORY OF SETS, E. Kamke. Clearest, amplest introduction in English, well suited for independent study. Subdivision of main theory, such as theory of sets of points, are discussed, but emphasis is on general theory. Partial contents: rudiments of set theory, arbitrary sets and their cardinal numbers, ordered sets and their order types, well-ordered sets and their cardinal numbers. Bibliography. Key to symbols. Index. vii + 144pp. 5⅜ x 8. S141 Paperbound **$1.35**

THEORY AND APPLICATIONS OF FINITE GROUPS, G. A. Miller, H. F. Blichfeldt, L. E. Dickson. Unusually accurate and authoritative work, each section prepared by a leading specialist: Miller on substitution and abstract groups, Blichfeldt on finite groups of linear homogeneous transformations, Dickson on applications of finite groups. Unlike more modern works, this gives the concrete basis from which abstract group theory arose. Includes Abelian groups, prime-power groups, isomorphisms, matrix forms of linear transformations, Sylow groups, Galois' theory of algebraic equations, duplication of a cube, trisection of an angle, etc. 2 Indexes. 267 problems. xvii + 390pp. 5⅜ x 8. S216 Paperbound **$2.00**

THE THEORY OF DETERMINANTS, MATRICES, AND INVARIANTS, H. W. Turnbull. Important study includes all salient features and major theories. 7 chapters on determinants and matrices cover fundamental properties, Laplace identities, multiplication, linear equations, rank and differentiation, etc. Sections on invariants gives general properties, symbolic and direct methods of reduction, binary and polar forms, general linear transformation, first fundamental theorem, multilinear forms. Following chapters study development and proof of Hilbert's Basis Theorem, Gordan-Hilbert Finiteness Theorem, Clebsch's Theorem, and include discussions of apolarity, canonical forms, geometrical interpretations of algebraic forms, complete system of the general quadric, etc. New preface and appendix. Bibliography. xviii + 374pp. 5⅜ x 8. S699 Paperbound **$2.25**

AN INTRODUCTION TO THE THEORY OF CANONICAL MATRICES, H. W. Turnbull and A. C. Aitken. All principal aspects of the theory of canonical matrices, from definitions and fundamental properties of matrices to the practical applications of their reduction to canonical form. Beginning with matrix multiplications, reciprocals, and partitioned matrices, the authors go on to elementary transformations and bilinear and quadratic forms. Also covers such topics as a rational canonical form for the collineatory group, congruent and conjunctive transformation for quadratic and hermitian forms, unitary and orthogonal transformations, canonical reduction of pencils of matrices, etc. Index. Appendix. Historical notes at chapter ends. Bibliographies. 275 problems. xiv + 200pp. 5⅜ x 8. S177 Paperbound **$1.55**

A TREATISE ON THE THEORY OF DETERMINANTS, T. Muir. Unequalled as an exhaustive compilation of nearly all the known facts about determinants up to the early 1930's. Covers notation and general properties, row and column transformation, symmetry, compound determinants, adjugates, rectangular arrays and matrices, linear dependence, gradients, Jacobians, Hessians, Wronskians, and much more. Invaluable for libraries of industrial and research organizations as well as for student, teacher, and mathematician; very useful in the field of computing machines. Revised and enlarged by W. H. Metzler. Index. 485 problems and scores of numerical examples. iv + 766pp. 5⅜ x 8. S670 Paperbound **$3.00**

THEORY OF DETERMINANTS IN THE HISTORICAL ORDER OF DEVELOPMENT, Sir Thomas Muir. Unabridged reprinting of this complete study of 1,859 papers on determinant theory written between 1693 and 1900. Most important and original sections reproduced, valuable commentary on each. No other work is necessary for determinant research: all types are covered—each subdivision of the theory treated separately; all papers dealing with each type are covered; you are told exactly what each paper is about and how important its contribution is. Each result, theory, extension, or modification is assigned its own identifying numeral so that the full history may be more easily followed. Includes papers on determinants in general, determinants and linear equations, symmetric determinants, alternants, recurrents, determinants having invariant factors, and all other major types. "A model of what such histories ought to be," NATURE. "Mathematicians must ever be grateful to Sir Thomas for his monumental work," AMERICAN MATH MONTHLY. Four volumes bound as two. Indices. Bibliographies. Total of lxxxiv + 1977pp. 5⅜ x 8. S672-3 The set, Clothbound **$12.50**

Calculus and function theory, Fourier theory, infinite series, calculus of variations, real and complex functions

FIVE VOLUME "THEORY OF FUNCTIONS' SET BY KONRAD KNOPP

This five-volume set, prepared by Konrad Knopp, provides a complete and readily followed account of theory of functions. Proofs are given concisely, yet without sacrifice of completeness or rigor. These volumes are used as texts by such universities as M.I.T., University of Chicago, N. Y. City College, and many others. "Excellent introduction . . . remarkably readable, concise, clear, rigorous," JOURNAL OF THE AMERICAN STATISTICAL ASSOCIATION.

ELEMENTS OF THE THEORY OF FUNCTIONS, Konrad Knopp. This book provides the student with background for further volumes in this set, or texts on a similar level. Partial contents: foundations, system of complex numbers and the Gaussian plane of numbers, Riemann sphere of numbers, mapping by linear functions, normal forms, the logarithm, the cyclometric functions and binomial series. "Not only for the young student, but also for the student who knows all about what is in it," MATHEMATICAL JOURNAL. Bibliography. Index. 140pp. 5⅜ x 8. S154 Paperbound **$1.50**

THEORY OF FUNCTIONS, PART I, Konrad Knopp. With volume II, this book provides coverage of basic concepts and theorems. Partial contents: numbers and points, functions of a complex variable, integral of a continuous function, Cauchy's integral theorem, Cauchy's integral formulae, series with variable terms, expansion of analytic functions in power series, analytic continuation and complete definition of analytic functions, entire transcendental functions, Laurent expansion, types of singularities. Bibliography. Index. vii + 146pp. 5⅜ x 8. S156 Paperbound **$1.35**

THEORY OF FUNCTIONS, PART II, Konrad Knopp. Application and further development of general theory, special topics. Single valued functions, entire, Weierstrass, Meromorphic functions. Riemann surfaces. Algebraic functions. Analytical configuration, Riemann surface. Bibliography. Index. x + 150pp. 5⅜ x 8. S157 Paperbound **$1.35**

PROBLEM BOOK IN THE THEORY OF FUNCTIONS, VOLUME 1, Konrad Knopp. Problems in elementary theory, for use with Knopp's THEORY OF FUNCTIONS, or any other text, arranged according to increasing difficulty. Fundamental concepts, sequences of numbers and infinite series, complex variable, integral theorems, development in series, conformal mapping. 182 problems. Answers. viii + 126pp. 5⅜ x 8. S158 Paperbound **$1.35**

PROBLEM BOOK IN THE THEORY OF FUNCTIONS, VOLUME 2, Konrad Knopp. Advanced theory of functions, to be used either with Knopp's THEORY OF FUNCTIONS, or any other comparable text. Singularities, entire & meromorphic functions, periodic, analytic, continuation, multiple-valued functions, Riemann surfaces, conformal mapping. Includes a section of additional elementary problems. "The difficult task of selecting from the immense material of the modern theory of functions the problems just within the reach of the beginner is here masterfully accomplished," AM. MATH. SOC. Answers. 138pp. 5⅜ x 8. S159 Paperbound **$1.35**

A COURSE IN MATHEMATICAL ANALYSIS, Edouard Goursat. Trans. by E. R. Hedrick, O. Dunkel. Classic study of fundamental material thoroughly treated. Exceptionally lucid exposition of wide range of subject matter for student with 1 year of calculus. Vol. 1: Derivatives and Differentials, Definite Integrals, Expansion in Series, Applications to Geometry. Problems. Index. 52 illus. 556pp. Vol. 2, Part I: Functions of a Complex Variable, Conformal Representations, Doubly Periodic Functions, Natural Boundaries, etc. Problems. Index. 38 illus. 269pp. Vol. 2, Part 2: Differential Equations, Cauchy-Lipschitz Method, Non-linear Differential Equations, Simultaneous Equations, etc. Problems. Index. 308pp. 5⅜ x 8.
Vol. 1 S554 Paperbound **$2.50**
Vol. 2 part 1 S555 Paperbound **$1.85**
Vol. 2 part 2 S556 Paperbound **$1.85**
3 vol. set **$6.20**

MODERN THEORIES OF INTEGRATION, H. Kestelman. Connected and concrete coverage, with fully-worked-out proofs for every step. Ranges from elementary definitions through theory of aggregates, sets of points, Riemann and Lebesgue integration, and much more. This new revised and enlarged edition contains a new chapter on Riemann-Stieltjes integration, as well as a supplementary section of 186 exercises. Ideal for the mathematician, student, teacher, or self-studier. Index of Definitions and Symbols. General Index. Bibliography. x + 310pp. 5⅝ x 8⅜. S572 Paperbound **$2.25**

THEORY OF MAXIMA AND MINIMA, H. Hancock. Fullest treatment ever written; only work in English with extended discussion of maxima and minima for functions of 1, 2, or n variables, problems with subsidiary constraints, and relevant quadratic forms. Detailed proof of each important theorem. Covers the Scheeffer and von Dantscher theories, homogeneous quadratic forms, reversion of series, fallacious establishment of maxima and minima, etc. Unsurpassed treatise for advanced students of calculus, mathematicians, economists, statisticians. Index. 24 diagrams. 39 problems, many examples. 193pp. 5⅜ x 8. S665 Paperbound **$1.50**

AN ELEMENTARY TREATISE ON ELLIPTIC FUNCTIONS, A. Cayley. Still the fullest and clearest text on the theories of Jacobi and Legendre for the advanced student (and an excellent supplement for the beginner). A masterpiece of exposition by the great 19th century British mathematician (creator of the theory of matrices and abstract geometry), it covers the addition-theory, Landen's theorem, the 3 kinds of elliptic integrals, transformations, the q-functions, reduction of a differential expression, and much more. Index. xii + 386pp. 5⅜ x 8.
S728 Paperbound **$2.00**

THE APPLICATIONS OF ELLIPTIC FUNCTIONS, A. G. Greenhill. Modern books forego detail for sake of brevity—this book offers complete exposition necessary for proper understanding, use of elliptic integrals. Formulas developed from definite physical, geometric problems; examples representative enough to offer basic information in widely useable form. Elliptic integrals, addition theorem, algebraical form of addition theorem, elliptic integrals of 2nd, 3rd kind, double periodicity, resolution into factors, series, transformation, etc. Introduction. Index. 25 illus. xi + 357pp. 5⅜ x 8. S603 Paperbound **$1.75**

THE THEORY OF FUNCTIONS OF REAL VARIABLES, James Pierpont. A 2-volume authoritative exposition, by one of the foremost mathematicians of his time. Each theorem stated with all conditions, then followed by proof. No need to go through complicated reasoning to discover conditions added without specific mention. Includes a particularly complete, rigorous presentation of theory of measure; and Pierpont's own work on a theory of Lebesgue integrals, and treatment of area of a curved surface. Partial contents, Vol. 1: rational numbers, exponentials, logarithms, point aggregates, maxima, minima, proper integrals, improper integrals, multiple proper integrals, continuity, discontinuity, indeterminate forms. Vol. 2: point sets, proper integrals, series, power series, aggregates, ordinal numbers, discontinuous functions, sub-, infra-uniform convergence, much more. Index. 95 illustrations. 1229pp. 5⅜ x 8. S558-9, 2 volume set, paperbound **$5.20**

FUNCTIONS OF A COMPLEX VARIABLE, James Pierpont. Long one of best in the field. A thorough treatment of fundamental elements, concepts, theorems. A complete study, rigorous, detailed, with carefully selected problems worked out to illustrate each topic. Partial contents: arithmetical operations, real term series, positive term series, exponential functions, integration, analytic functions, asymptotic expansions, functions of Weierstrass, Legendre, etc. Index. List of symbols. 122 illus. 597pp. 5⅜ x 8. S560 Paperbound **$2.45**

MODERN OPERATIONAL CALCULUS: WITH APPLICATIONS IN TECHNICAL MATHEMATICS, N. W. McLachlan. An introduction to modern operational calculus based upon the Laplace transform, applying it to the solution of ordinary and partial differential equations. For physicists, engineers, and applied mathematicians. Partial contents: Laplace transform, theorems or rules of the operational calculus, solution of ordinary and partial linear differential equations with constant coefficients, evaluation of integrals and establishment of mathematical relationships, derivation of Laplace transforms of various functions, etc. Six appendices deal with Heaviside's unit function, etc. Revised edition. Index. Bibliography. xiv + 218pp. 5⅜ x 8½. S192 Paperbound **$1.75**

ADVANCED CALCULUS, E. B. Wilson. An unabridged reprinting of the work which continues to be recognized as one of the most comprehensive and useful texts in the field. It contains an immense amount of well-presented, fundamental material, including chapters on vector functions, ordinary differential equations, special functions, calculus of variations, etc., which are excellent introductions to these areas. For students with only one year of calculus, more than 1300 exercises cover both pure math and applications to engineering and physical problems. For engineers, physicists, etc., this work, with its 54 page introductory review, is the ideal reference and refresher. Index. ix + 566pp. 5⅜ x 8. S504 Paperbound **$2.45**

ASYMPTOTIC EXPANSIONS, A. Erdélyi. The only modern work available in English, this is an unabridged reproduction of a monograph prepared for the Office of Naval Research. It discusses various procedures for asymptotic evaluation of integrals containing a large parameter and solutions of ordinary linear differential equations. Bibliography of 71 items. vi + 108pp. 5⅜ x 8. S318 Paperbound **$1.35**

INTRODUCTION TO ELLIPTIC FUNCTIONS: with applications, F. Bowman. Concise, practical introduction to elliptic integrals and functions. Beginning with the familiar trigonometric functions, it requires nothing more from the reader than a knowledge of basic principles of differentiation and integration. Discussion confined to the Jacobian functions. Enlarged bibliography. Index. 173 problems and examples. 56 figures, 4 tables. 115pp. 5⅜ x 8. S922 Paperbound **$1.50**

ON RIEMANN'S THEORY OF ALGEBRAIC FUNCTIONS AND THEIR INTEGRALS: A SUPPLEMENT TO THE USUAL TREATISES, Felix Klein. Klein demonstrates how the mathematical ideas in Riemann's work on Abelian integrals can be arrived at by thinking in terms of the flow of electric current on surfaces. Intuitive explanations, not detailed proofs given in an extremely clear exposition, concentrating on the kinds of functions which can be defined on Riemann surfaces. Also useful as an introduction to the origins of topological problems. Complete and unabridged. Approved translation by Frances Hardcastle. New introduction. 43 figures. Glossary. xii + 76pp. 5⅜ x 8½. S1072 Paperbound **$1.25**

COLLECTED WORKS OF BERNHARD RIEMANN. This important source book is the first to contain the complete text of both 1892 Werke and the 1902 supplement, unabridged. It contains 31 monographs, 3 complete lecture courses, 15 miscellaneous papers, which have been of enormous importance in relativity, topology, theory of complex variables, and other areas of mathematics. Edited by R. Dedekind, H. Weber, M. Noether, W. Wirtinger. German text. English introduction by Hans Lewy. 690pp. 5⅜ x 8. S226 Paperbound **$3.75**

THE TAYLOR SERIES, AN INTRODUCTION TO THE THEORY OF FUNCTIONS OF A COMPLEX VARIABLE, P. Dienes. This book investigates the entire realm of analytic functions. Only ordinary calculus is needed, except in the last two chapters. Starting with an introduction to real variables and complex algebra, the properties of infinite series, elementary functions, complex differentiation and integration are carefully derived. Also biuniform mapping, a thorough two part discussion of representation and singularities of analytic functions, overconvergence and gap theorems, divergent series, Taylor series on its circle of convergence, divergence and singularities, etc. Unabridged, corrected reissue of first edition. Preface and index. 186 examples, many fully worked out. 67 figures. xii + 555pp. 5⅜ x 8. S391 Paperbound **$2.75**

INTRODUCTION TO BESSEL FUNCTIONS, Frank Bowman. A rigorous self-contained exposition providing all necessary material during the development, which requires only some knowledge of calculus and acquaintance with differential equations. A balanced presentation including applications and practical use. Discusses Bessel Functions of Zero Order, of Any Real Order; Modified Bessel Functions of Zero Order; Definite Integrals; Asymptotic Expansions; Bessel's Solution to Kepler's Problem; Circular Membranes; much more. "Clear and straightforward . . . useful not only to students of physics and engineering, but to mathematical students in general," Nature. 226 problems. Short tables of Bessel functions. 27 figures. Index. x + 135pp. 5⅜ x 8. S462 Paperbound **$1.50**

ELEMENTS OF THE THEORY OF REAL FUNCTIONS, J. E. Littlewood. Based on lectures given at Trinity College, Cambridge, this book has proved to be extremely successful in introducing graduate students to the modern theory of functions. It offers a full and concise coverage of classes and cardinal numbers, well-ordered series, other types of series, and elements of the theory of sets of points. 3rd revised edition. vii + 71pp. 5⅜ x 8.
S171 Clothbound **$2.85**
S172 Paperbound **$1.25**

TRANSCENDENTAL AND ALGEBRAIC NUMBERS, A. O. Gelfond. First English translation of work by leading Soviet mathematician. Thue-Siegel theorem, its p-adic analogue, on approximation of algebraic numbers by numbers in fixed algebraic field; Hermite-Lindemann theorem on transcendency of Bessel functions, solutions of other differential equations; Gelfond-Schneider theorem on transcendency of alpha to power beta; Schneider's work on elliptic functions, with method developed by Gelfond. Translated by L. F. Boron. Index. Bibliography. 200pp. 5⅜ x 8. S615 Paperbound **$1.75**

ELLIPTIC INTEGRALS, H. Hancock. Invaluable in work involving differential equations containing cubics or quartics under the root sign, where elementary calculus methods are inadequate. Practical solutions to problems that occur in mathematics, engineering, physics: differential equations requiring integration of Lamé's, Briot's, or Bouquet's equations; determination of arc of ellipse, hyperbola, lemniscate; solutions of problems in elastica; motion of a projectile under resistance varying as the cube of the velocity; pendulums; many others. Exposition is in accordance with Legendre-Jacobi theory and includes rigorous discussion of Legendre transformations. 20 figures. 5 place table. Index. 104pp. 5⅛ x 8.
S484 Paperbound **$1.25**

LECTURES ON THE THEORY OF ELLIPTIC FUNCTIONS, H. Hancock. Reissue of the only book in English with so extensive a coverage, especially of Abel, Jacobi, Legendre, Weierstrasse, Hermite, Liouville, and Riemann. Unusual fullness of treatment, plus applications as well as theory, in discussing elliptic function (the universe of elliptic integrals originating in works of Abel and Jacobi), their existence, and ultimate meaning. Use is made of Riemann to provide the most general theory. 40 page table of formulas. 76 figures. xxiii + 498pp.
S483 Paperbound **$2.55**

THE THEORY AND FUNCTIONS OF A REAL VARIABLE AND THE THEORY OF FOURIER'S SERIES, E. W. Hobson. One of the best introductions to set theory and various aspects of functions and Fourier's series. Requires only a good background in calculus. Provides an exhaustive coverage of: metric and descriptive properties of sets of points; transfinite numbers and order types; functions of a real variable; the Riemann and Lebesgue integrals; sequences and series of numbers; power-series; functions representable by series sequences of continuous functions; trigonometrical series; representation of functions by Fourier's series; complete exposition (200pp.) on set theory; and much more. "The best possible guide," Nature. Vol. I: 88 detailed examples, 10 figures. Index. xv + 736pp. Vol. II: 117 detailed examples, 13 figures. Index. x + 780pp. 6⅛ x 9¼. Vol. I: S387 Paperbound **$3.50**
Vol. II: S388 Paperbound **$3.00**

ALMOST PERIODIC FUNCTIONS, A. S. Besicovitch. This unique and important summary by a well-known mathematician covers in detail the two stages of development in Bohr's theory of almost periodic functions: (1) as a generalization of pure periodicity, with results and proofs; (2) the work done by Stepanoff, Wiener, Weyl, and Bohr in generalizing the theory. Bibliography. xi + 180pp. 5⅜ x 8. S18 Paperbound **$1.75**

THE ANALYTICAL THEORY OF HEAT, Joseph Fourier. This book, which revolutionized mathematical physics, is listed in the Great Books program, and many other listings of great books. It has been used with profit by generations of mathematicians and physicists who are interested in either heat or in the application of the Fourier integral. Covers cause and reflection of rays of heat, radiant heating, heating of closed spaces, use of trigonometric series in the theory of heat, Fourier integral, etc. Translated by Alexander Freeman. 20 figures. xxii + 466pp. 5⅜ x 8. S93 Paperbound **$2.50**

AN INTRODUCTION TO FOURIER METHODS AND THE LAPLACE TRANSFORMATION, Philip Franklin. Concentrates upon essentials, enabling the reader with only a working knowledge of calculus to gain an understanding of Fourier methods in a broad sense, suitable for most applications. This work covers complex qualities with methods of computing elementary functions for complex values of the argument and finding approximations by the use of charts; Fourier series and integrals with half-range and complex Fourier series; harmonic analysis; Fourier and Laplace transformations, etc.; partial differential equations with applications to transmission of electricity; etc. The methods developed are related to physical problems of heat flow, vibrations, electrical transmission, electromagnetic radiation, etc. 828 problems with answers. Formerly entitled "Fourier Methods." Bibliography. Index. x + 289pp. 5⅜ x 8.
S452 Paperbound **$2.00**

THE FOURIER INTEGRAL AND CERTAIN OF ITS APPLICATIONS, Norbert Wiener. The only book-length study of the Fourier integral as link between pure and applied math. An expansion of lectures given at Cambridge. Partial contents: Plancherel's theorem, general Tauberian theorem, special Tauberian theorems, generalized harmonic analysis. Bibliography. viii + 201pp. 5⅜ x 8. S272 Paperbound **$1.50**

INTRODUCTION TO THE THEORY OF FOURIER'S SERIES AND INTEGRALS, H. S. Carslaw. 3rd revised edition. This excellent introduction is an outgrowth of the author's courses at Cambridge. Historical introduction, rational and irrational numbers, infinite sequences and series, functions of a single variable, definite integral, Fourier series, Fourier integrals, and similar topics. Appendixes discuss practical harmonic analysis, periodogram analysis. Lebesgue's theory. Indexes. 84 examples, bibliography. xii + 368pp. 5⅜ x 8. S48 Paperbound **$2.00**

FOURIER'S SERIES AND SPHERICAL HARMONICS, W. E. Byerly. Continues to be recognized as one of most practical, useful expositions. Functions, series, and their differential equations are concretely explained in great detail; theory is applied constantly to practical problems, which are fully and lucidly worked out. Appendix includes 6 tables of surface zonal harmonics, hyperbolic functions, Bessel's functions. Bibliography. 190 problems, approximately half with answers. ix + 287pp. 5⅜ x 8. S536 Paperbound **$1.75**

INFINITE SEQUENCES AND SERIES, Konrad Knopp. First publication in any language! Excellent introduction to 2 topics of modern mathematics, designed to give the student background to penetrate farther by himself. Sequences & sets, real & complex numbers, etc. Functions of a real & complex variable. Sequences & series. Infinite series. Convergent power series. Expansion of elementary functions. Numerical evaluation of series. Bibliography. v + 186pp. 5⅜ x 8. S153 Paperbound **$1.75**

TRIGONOMETRICAL SERIES, Antoni Zygmund. Unique in any language on modern advanced level. Contains carefully organized analyses of trigonometric, orthogonal, Fourier systems of functions, with clear adequate descriptions of summability of Fourier series, proximation theory, conjugate series, convergence, divergence of Fourier series. Especially valuable for Russian, Eastern European coverage. Bibliography. 329pp. 5⅜ x 8. S290 Paperbound **$1.50**

DICTIONARY OF CONFORMAL REPRESENTATIONS, H. Kober. Laplace's equation in 2 dimensions solved in this unique book developed by the British Admiralty. Scores of geometrical forms & their transformations for electrical engineers, Joukowski aerofoil for aerodynamists. Schwarz-Christoffel transformations for hydrodynamics, transcendental functions. Contents classified according to analytical functions describing transformation. Twin diagrams show curves of most transformations with corresponding regions. Glossary. Topological index. 447 diagrams. 244pp. 6⅛ x 9¼. S160 Paperbound **$2.00**

CALCULUS OF VARIATIONS, A. R. Forsyth. Methods, solutions, rather than determination of weakest valid hypotheses. Over 150 examples completely worked-out show use of Euler, Legendre, Jacobi, Weierstrass tests for maxima, minima. Integrals with one original dependent variable; with derivatives of 2nd order; two dependent variables, one independent variable; double integrals involving 1 dependent variable, 2 first derivatives; double integrals involving partial derivatives of 2nd order; triple integrals; much more. 50 diagrams. 678pp. 5⅝ x 8⅜. S622 Paperbound **$2.95**

LECTURES ON THE CALCULUS OF VARIATIONS, O. Bolza. Analyzes in detail the fundamental concepts of the calculus of variations, as developed from Euler to Hilbert, with sharp formulations of the problems and rigorous demonstrations of their solutions. More than a score of solved examples; systematic references for each theorem. Covers the necessary and sufficient conditions; the contributions made by Euler, Du Bois-Reymond, Hilbert, Weierstrass, Legendre, Jacobi, Erdmann, Kneser, and Gauss; and much more. Index. Bibliography. xi + 271pp. 5⅜ x 8. S218 Paperbound **$1.65**

A TREATISE ON THE CALCULUS OF FINITE DIFFERENCES, G. Boole. A classic in the literature of the calculus. Thorough, clear discussion of basic principles, theorems, methods. Covers MacLaurin's and Herschel's theorems, mechanical quadrature, factorials, periodical constants, Bernoulli's numbers, difference-equations (linear, mixed, and partial), etc. Stresses analogies with differential calculus. 236 problems, answers to the numerical ones. viii + 336pp. 5⅜ x 8. S695 Paperbound **$1.85**

Prices subject to change without notice.

Dover publishes books on art, music, philosophy, literature, languages, history, social sciences, psychology, handcrafts, orientalia, puzzles and entertainments, chess, pets and gardens, books explaining science, intermediate and higher mathematics, mathematical physics, engineering, biological sciences, earth sciences, classics of science, etc. Write to:

Dept. catrr.
Dover Publications, Inc.
180 Varick Street, N.Y. 14, N.Y.